CW00506885

the
Regenerative
Garden

80 PRACTICAL PROJECTS
FOR CREATING A SELF-SUSTAINING
GARDEN ECOSYSTEM

Stephanie Rose

Inspiring | Educating | Creating | Entertaining

Brimming with creative inspiration, how-to projects, and useful information to enrich your everyday life, Quarto.com is a favorite destination for those pursuing their interests and passions.

© 2022 Quarto Publishing Group USA Inc.
Text and photography © 2022 Stephanie Rose

First Published in 2022 by Cool Springs Press, an imprint of The Quarto Group, 100 Cummings Center, Suite 265-D, Beverly, MA 01915, USA.
T (978) 282-9590 F (978) 283-2742 Quarto.com

All rights reserved. No part of this book may be reproduced in any form without written permission of the copyright owners. All images in this book have been reproduced with the knowledge and prior consent of the artists concerned, and no responsibility is accepted by producer, publisher, or printer for any infringement of copyright or otherwise, arising from the contents of this publication. Every effort has been made to ensure that credits accurately comply with information supplied. We apologize for any inaccuracies that may have occurred and will resolve inaccurate or missing information in a subsequent reprinting of the book.

Cool Springs Press titles are also available at discount for retail, wholesale, promotional, and bulk purchase. For details, contact the Special Sales Manager by email at specialsales@quarto.com or by mail at The Quarto Group, Attn: Special Sales Manager, 100 Cummings Center, Suite 265-D, Beverly, MA 01915, USA.

26 25 24 23 22   1 2 3 4 5

ISBN: 978-0-7603-7168-8

Digital edition published in 2022
eISBN: 978-0-7603-7170-1

Library of Congress Cataloging-in-Publication Data available

Design: Allison Meierding
Page Layout: Allison Meierding
Photography: Stephanie Rose except pages 51, 52, 53, 54 Steven Biggs; 71 Lori Weidenhammer; 104, 105 Harris Seeds; 110 (bottom) Tanya Anderson; 112, 113, 114, 115 (top left) Michael Rose; 149, 150, 151, 152 Holly Rodgers; 13, 14 (bottom right), 36 (left middle), 38, 42, 60 (bottom left), 86 (bottom), 94 (bottom right), 98, 118 (bottom left), 127, 136 (bottom left), 142 (bottom), 169 Eduardo Cristo; 7 (top left), 115 (bottom) Susan Goble / Wildflower Photography, 103 James Miceli; 133 (top) Melody Kurt
Illustration: Ada Keesler @adagracee except page 28 Mattie Wells

Printed in China

Dedication

⬤ ⬤ ⬤ ⬤ ⬤ ⬤

To Asher, Protector of Bees, Grazer of Flowers, and budding Eco-Activist.
Keep on growing your passion for the natural world. Stand tall and
speak loudly. I'll be right there beside you along the way.

Contents

The Regenerative Garden
Project List

Chapter 1 |
Soil Growing from
the Ground Up

Chapter 2 |
Water Efficiency in
Collection and Use

Chapter 3 |
Plants Growing Life

Chapter 4 | Climate
Creating Harmony

Chapter 5 | Ethics
Reducing Waste &
Encouraging Diversity

Chapter 6 | Community
Building Sharing Spaces for
Everyone

Foreword

This may be the only gardening book you'll ever need! In her brilliantly simple yet profound manner, Stephanie Rose guides us through the steps of creating a garden that is not only beautiful and easy to maintain, but also sustainable and regenerative. Based on the ethics or principles of permaculture—*Earth Share, People Share, and Fair Share*—regenerative gardening is gardening the way Mother Nature intended. We are assisting nature in her job and reaping many benefits as we do so. You may wonder how easy this is to do in our backyards and gardens while still maintaining some sense of "gardening order" and design. That's precisely what Stephanie sets out to show us in the pages of this book, and she does so in simple, practical, and easy-to-do steps.

Stephanie is a competent gardening guide and has designed and created her own home and several community regenerative gardens. As you'll read in her opening remarks, "My garden feeds my family and gives us a place to play without the stresses of weeding, pest control, or endless watering. My hope is that you can enjoy this same gift by building your own regenerative garden." Imagine a garden space that invites pollinators, restores soil, creates healthy ecosystems rich with diversity, and connects neighbors and communities, all while requiring minimum maintenance or upkeep—that's regenerative gardening!

Although the concepts of permaculture and regenerative gardening can sometimes seem overwhelming to those who have been using more conventional methods to grow, Stephanie is not only a skilled gardener, but also an excellent teacher and writer. From soil amendments, watering systems, polyculture, and garden design to creating climate harmony in our gardens, Stephanie walks us through each step with easy, cost-effective projects designed for the home gardener. Her simple but detailed explanations and commonsense directions make regenerative gardening an exciting possibility in any home setting. As she states, "the goal is not perfection but progress" and goes on to add, "If one can apply any of the concepts in this book to your home, garden, or life—then be proud!" This is how we'll change the world into the garden paradise we dream of—one regenerative garden at a time.

—Rosemary Gladstar
Herbalist & Author
From her gardens at Misty Bay, Vermont

Introduction

Imagine

Imagine your garden today. In your mind's eye, walk through the gate or out the door into the garden space. Look around and take in how she looks at this moment.

    Now imagine that you've suddenly been called away and have to leave your garden immediately. There will be no one to water, fertilize, or weed until you return. The garden must fend for herself.

One Year Later

A year has passed and you are finally able to return to your garden again. In your mind's eye, walk into the garden, look around, and observe how she looks in this moment.

When I do this exercise in my live talks, I ask the audience to call out what the garden looks like after being entirely neglected by humans for an entire year. The usual responses are that the garden is dried out, dead, dying, pest-ridden, or diseased. In some areas, the soil has become barren and in others it has become overgrown with weeds.

Ten Years Later

Now imagine you are not able to return to the garden for ten years. No human has maintained the garden space for an entire decade. In your mind's eye, walk into the garden, look around, and observe how she looks in this moment.

When I ask this question, I get completely different replies from the audience. Folks call out that the garden is lush and thriving, full of wildlife, and bursting with plants. She has returned to her natural ecosystem. The plants and animals that thrive without human input have moved in and taken over.

This is a regenerative garden.

Regenerative is beyond the concept of sustainability. It's beyond the practice of resiliency. It's a space that can thrive on its own without human input, as Mother Nature has been doing since the beginning.

Doesn't it make sense to create this kind of ecosystem in our home gardens? We can reduce the efforts we put into managing our gardens and allow nature to guide us. This is not to say that our gardens must become wild and unmanicured. We can implement systems and projects to build modern-day regenerative spaces that follow the guidance of nature and the designs that integrate our small patches of landscape with our homes and community. Through building the soil, choosing biodiverse plants, caching and storing water, working with microclimates, reusing waste, and connecting with our neighbors, we can design a garden that regenerates itself and feeds us.

Good. Better. Best.

Permaculture teacher Hazel Ward[1] presented the concept of "good, better, best" when it comes to design—meaning there are many levels of benefits that can be gained for the earth, our communities, and our gardens through permaculture designs, but that there are many factors that influence what we can accomplish. In reality, we all have varying access to time, materials, skills, and money, and oftentimes it takes many steps to get to where we want to be. When designing projects for your garden, "good" could look like a quick fix that improves sustainability; "better" could add some resiliency to adapt better with less input; and "best" would operate as a self-sustaining and regenerative garden.

That being said, the worst-case scenario would be to for me to present ideas for regenerative design that are so expensive and out of reach that very few people can achieve them. Instead, I'll explain the concepts

PERMACULTURE ETHICS

Permaculture is guided by three main ethics: Earth Care, People Care, and Fair Share. Recently, a fourth ethic has become better known, the Transition Ethic, which recognizes that a long journey must start with smaller steps. Keeping this in mind, each project and chapter introduction is labeled with one to three honeycomb links that represent the Transition Ethic. The first link represents a place to start and the other links are ideas that build even more regenerative home landscapes.

I encourage you to strive for doing little bits of "good" by trying the projects that resonate with you. Soon you should start to feel the rewards of the work you're doing to create a regenerative space. You'll be doing fewer garden chores, saving money, growing healthier plants, and enjoying a more fruitful harvest. Once you see the results, I know you'll bring this book back out year after year and add more projects to your garden space, moving toward creating a self-sustaining home ecosystem.

and give you the steps for projects you can do easily and cost effectively, with ideas and suggestions for how to take it further if you wish to. These concepts can be applied and inputs adjusted in order to make some changes to our modern-day unsustainable garden designs so that we begin to move in the right direction.

It's with this concept in mind that I've designed the projects for this book. Each project is listed with some notes for how it can be applied at home as good, better, or ~~best~~ even better. There's no grading system because any steps toward regeneration are the right steps. At times, unsustainable methods or materials must be used in order to transition to a more regenerative place. The goal here is not perfection, it's progress. If you can apply any of the concepts in this book to your home, garden, or life—then be proud! All the changes we make at home have a great impact on our global well-being.

Every year that I get to enjoy a lush garden space that takes only minimal human input, I'm grateful to have designed a regenerative garden. My guests are surprised by how little work it is to have a garden jam-packed with plants, water gardens, and wildlife. My garden feeds my family and gives us a place to play without the stresses of weeding, pest control, or endless watering. My hope is that you can enjoy this same gift by building your own regenerative garden.

Stephanie

1

Soil

Growing from the Ground Up

Feed and nourish
your soil to grow
better plants.

Recycle waste and
plant soil-building
amendments.

Learn to "read the
weeds" and plant soil
fixers to regenerate soil.

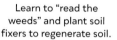

Think of the ocean. It's a vast ecosystem inhabited by a variety of creatures that live within it, some of which we have met and others that we have yet to meet. It's both awe-inspiring and mysterious how it exists and how it gives life. This could also be an accurate description of the soil beneath our feet. Those of us who have gotten to know soil respect it for the extensive ecosystem that lives below its surface. Bustling activity goes on—day in and day out—to break down organic matter into nutrients available for plant uptake. It cycles and works together, communicating needs in a language that we are just beginning to understand.

Soil is an ancient system that can outsmart any modern-day technology. Plants naturally grow together that are needed to support one another. As leaves fall, they compost and break down onto the surface level to feed the young, shallowly rooted plants. As these plants grow older, they mine down for minerals deep in the soil, bringing them up through their stems, leaves, and flowers and then drop them again onto the soil surface to be broken down by the microorganisms and insects that devour plant material and leave behind plant-available minerals and nutrients. Brilliant.

This soil-smart system is also self-regulating. You can't overfertilize because the plants only take up what they need and leave the rest stored in the soil for future use. If nutrients are missing the plants will die and new plants will show up in their place. Soil organisms have a predator-and-prey cycle that helps regulate their populations as well. If there's too much food available, their population will increase; if there's too little food, their population will decrease. This self-regulating smart system runs without any human input. The more we try to control it, the more we dumb it down.

To develop this sort of technology in our home gardens, all we need to do is start feeding the soil naturally with organic matter, compost, and mulch. Then step back and let it do the things it's supposed to do. The soil wildlife will move in and maintain it for us. This chapter covers a number of projects that will help reboot your garden from the ground up.

How to Amend Soil with Manure

- Add fully composted manure to your gardens annually.
- Use fully composted manure from local animals. Avoid manure from cats and dogs, which can contain pathogens.
- Match fully composted, locally sourced manure to plant type.

Manure is a common organic fertilizer that is inexpensive and readily available. Agriculture waste from animal or mushroom farming is composted and finished in order to be added as a natural fertilizer to your garden beds. That being said, not all manure is the same, and it should be used with caution.

The best practice is to compost manure completely before using it in the garden, although some gardeners and farmers do use fresh manure with success. Fresh manure is higher in nutrients, but it can also contain harmful bacteria and microorganisms.

The number of nutrients as well as the breakdown of each nutrient vary by manure type, age, input materials, and how it was composted. The ideal practice is to compost the manure from your own animals as they will be feeding on plants that are available in your property. If you don't have animals but want to add a rich source of fertilizer to the gardens, then sourcing manure may be a good option for you.

Composted manure is a safer option that allows the manure to "mellow"; composting can also sterilize undigested weed seeds and kill pathogens[2]. Compost manure in a compost pile or bin just as you would compost any organic matter. Choose a spot with good drainage where leachate (the liquid that is released from the compost pile) will not run into waterways. Manure needs to get hot enough to kill pathogens and weed seeds, so it's usually only done in large quantities such as a cubic yard minimum. After 24 to 48 hours, measure the temperature, which should reach 130 to 140 degrees F (54 to 60 degrees C). Turn the pile every one to three weeks to aerate, and water the pile in hot, dry conditions. When the manure has finished composting, the pile will cool down. At this stage, it needs to "cure" (age) for up to six months[3].

When composted manure has aged, it can be applied in spring and fall to garden beds by topdressing to ¼- to ½-inch (.6 to 1 cm) thickness, or by sidedressing established trees and shrubs at the same rate.

MANURE'S USES, BENEFITS, AND WARNINGS [4]

Manure	Best Used For	Benefits	Warnings
Chicken	Flowers and vegetables	Very high in nitrogen. Weed seeds destroyed when digested.	Must be composted. Too strong for direct application.
Cow/Steer	Flowers, vegetables, and acid-loving plants	Easily composted.	Smelly unless thoroughly composted.
Horse	Flowers, vegetables, and acid-loving plants	Rich in organic matter. Often includes straw.	Can be filled with undigested herbicides such as Clopyralid and Aminopyralid that will harm plants. Use with caution. [5]
Sheep/Llama	Topdress garden plants	Well digested and high in potassium. Slow-release fertilizer.	Labor-intensive to collect manure from field grazers.
Rabbit	Mulch for flowers, vegetables, and roses. Pellets can be used directly on lawns and to mulch roses, vegetables, and flower beds.	Pellets can be used fresh from the hutch. High in nitrogen and phosphorus. Can be used to speed up composting.	Wash your hands thoroughly after handling rabbit manure.
Mushroom	Ornamental gardens, trees, and shrubs	Perfect for plants that require a lot of water as it retains moisture.	Usually contains animal byproducts in the composted substrate. Not considered vegan.

How to Amend Soil with Compost

Compost is garden gold. It adds organic matter and microorganisms that work together to improve your soil. The larger particles of organic matter in compost create air pockets, hold moisture, and decompose into nutrients.

The soil organisms, such as insects, bacteria, and fungi, further break down the organic matter into nutrients that are available in a form that the plants can then use. The best practice in a regenerative garden is to create your own compost—not only to use the waste you create as a soil amendment, but because hyperlocal compost will naturally have more of the specific nutrients and organisms that your plants and soil need.

Amend soil with compost annually or more often when you have compost readily available. Spread compost in a thick layer over the soil when the garden is not productive. Compost does not need to be scratched in or turned into the top layer of the soil. Turning the compost in only disturbs the network of microorganisms already living in the soil. Better yet, add compost directly over the mulch layer that is protecting soil and decomposing itself (mulch layers are discussed later in this chapter).

- Topdress soil with compost annually.

- Use hyperlocal, homemade compost.

- Add seasonal layers of organic mulch and homemade compost.

WHAT IF YOU CAN'T MAKE ENOUGH COMPOST?

Often it is difficult to produce enough compost in a home garden to have enough to amend the soil. In this case you can purchase compost and mix it with your homemade compost. Buy the compost that you need and set it out on your property. Compost should not smell like sulfur or ammonia; it should have a sweet, earthy aroma. If the compost that's delivered is steaming and or smelly, allow it to cool down first. When the smell has gone, mix in as much of your own compost as you have available and apply it to your garden beds.

THE REGENERATIVE GARDEN

How to Amend Soil with Green Manures

- Grow green manures to condition and add nutrients to soil.

- Grow a biodiverse mix of green manure to chop and drop.

- Allow green manure to go to seed before cutting; harvest and save seeds.

Green manures, also called cover crops, are plants whose purpose is to be grown and then incorporated back into the soil to improve soil structure and add nutrients. Cover cropping is an effective way of inexpensively adding organic material and nutrients into the soil, particularly when using nitrogen-fixing crops. This practice has traditionally been done by growing the crops, cutting them down before flowers or seeds form, and then tilling the cut material into the soil. Today we have become more aware of the importance of soil structure and how tilling the soil destroys the delicate ecosystem.

In a regenerative garden, green manures are *not* tilled into the soil; instead, the plants are pulled or slashed, chopped into smaller pieces, and used as mulch on the soil surface. Below ground, the remaining roots can be left to decompose, which add valuable organic material. The mulch will decompose and provide nutrients and soil-conditioning effects while also retaining moisture, suppressing weeds, and maintaining the soil ecosystem.

Monocropping green manures doesn't give the soil as many of the benefits as a biodiverse mix of cover crops will. Combine nitrogen-fixing legumes with shallow roots with plants that have deeper roots such as plants with taproots or grasses.

Green manures that are grown longer, until they seed, allow more nutrient uptake to the plant as plants release chemicals into the soil to allow maximum nutrient absorption when they are setting seeds. Additionally, allowing plants to go to seed allows us to harvest that seed and use it for future crops. The key is to collect the seed as it ripens but before it falls and self-seeds.

Some crops planted in cooler climates are killed by frost, which offers a low-maintenance way to cut back the crop and return the organic material and nutrients to the soil. Note the hardiness temperature in the following chart to determine which might be appropriate for your climate.

When planting time comes, the mulch layer can be left in place and pulled back to plant crops. To speed the decomposition of the green manure and further boost soil health, the green manure can be topped with a layer of finished compost and another mulch layer of your choice.

Legumes such as lupins (*Lupinis* spp.) offer many benefits such as nitrogen-fixing, erosion control, and attracting pollinators.

SOIL

19

Charlie Nardozzi, author of *The Complete Guide to No-Dig Gardening*[6], completed a helpful guide to green manures for the National Gardening Bureau. It includes hardy legumes, tropical legumes, grasses, an other annual crops, to which I have added a few of my favorite green manures.

Hardy legumes are grown in cool climates as they are rich in nitrogen and organic matter and can handle cooler temperatures. They are planted in fall and then grow rapidly in early spring.

Tropical legumes need warm growing conditions in order to grow quickly in fall when planted in late summer or early fall (before winter cover crops). These can also be grown as summer annuals in cooler climates where the first freezes will naturally return the organic matter to the soil.

Grasses and other annual crops grow quickly, tolerate cold, increase organic matter, and improve the structure of compacted soils. They also control erosion but don't increase the amount of nitrogen.

Phacelia tanacetifolia is an easy-to-grow cover crop that attracts bees with its flowers and improves soil structure with its roots.

COMMON GREEN MANURES

Plant	Botanical Name	Height	Hardy to	Amount to Sow Per 1,000 sq.ft. (93 sq. meters)	Notes
LEGUMES					
Berseem Clover	*Trifolium alexandrinum*	1 to 2 feet (30 to 60 cm)	20°F (-6.6°C)	2 pounds (910 grams)	Matures late and fixes less nitrogen than other clovers. Attracts bees.
Crimson Clover	*Trifolium incarnatum*	18 inches (45 cm)	10°F (-12°C)	½ to 2 pounds (225 to 910 grams)	If allowed to go to seed, it can become weedy.
Dutch White Clover	*Trifolium repens*	6 to 8 inches (15 to 20 cm)	-20°F (-29°C)	½ to 1 pound (225 to 455 grams)	Can also be a long-term groundcover as it tolerates foot traffic well.
Alsike Clover	*Trifolium hybridum*	16 inches (40 cm)	-30°F (-34°C)	½ pound (225 grams)	Open-pollinated species native to Europe. Grows where many other clovers will not. May become weedy if not managed. Do not use in grazing pastures; toxic to horses.
Hairy Vetch	*Vicia villosa*	2 feet (60 cm)	-15°F (-26°C)	1 to 2 pounds (455 to 910 grams)	Hardiest annual legume. Tolerates poor soil, matures late.
Fava Bean	*Vicia faba*	3 to 8 feet (90 to 244 cm)	15°F (-9°C)	2 to 5 pounds (910 to 2,268 grams)	Bell bean is a shorter (3-foot) relative
Lupin	*Lupinus* spp.	3 feet (90 cm)	-20°F (-29°C)	1 to 1½ pounds (455 to 680 grams)	Nitrogen fixer, hay source, and grazing potential; erosion control; attracts pollinators; poor disease resistance[7].
Winter Field Peas	*Pisum sativum*	6 feet (183 cm)	-10°F (-23°C)	2 ½ pounds (1,135 grams)	Nitrogen-fixer, tolerates heavier soils. Combine with winter wheat for best results.
TROPICAL LEGUMES					
Sunn Hemp	*Crotolaria juncea*	5 to 6 feet (152 to 183 cm)	28°F (-2°C)	1 to 2 pounds (455 to 910 grams)	Needs same growing conditions as corn. Cut or mow before stems become woody. Can also reduce nematodes.
Sesbania	*Sesbania macrocarpa*	6 to 8 feet (183 to 244 cm)	32°F (0°C)	1 pound (455 grams)	Grows like sunn hemp but more tolerant of floods, drought, salinity, low fertility.
Cowpea	*Vigna sinensis*	1 to 2 feet (30 to 60 cm)	32°F (0°C)	2 to 3 pounds (910 to 1,361 grams)	Tolerates poor and acidic soils. Prefers humidity and tolerates drought.
Mung Bean	*Vigna radiata*	2 to 3 feet (60 to 90 cm)	32°F (0°C)	1 to 2 pounds (455 to 910 grams)	Fast to mature, heat- and drought-tolerant, grows in a variety of soils.[8]

Plant	Botanical Name	Height	Hardy to	Amount to Sow Per 1,000 sq.ft. (93 sq. meters)	Notes
GRASSES AND OTHER ANNUALS					
Oats	*Avena sativa*	2 to 3 feet (60 to 90 cm)	10 to 20°F (-18 to -6.6°C)	2 pounds (910 grams)	Produces least organic matter of grasses, but tolerant of wet soils.
Barley	*Hordeum vulgare*	2 to 3 feet (60 to 90 cm)	0 to 10 F (-18 to -12°C)	2 to 3 pounds (910 to 1,361 grams)	Fast maturing and tolerant of dry and saline soils; intolerant of acidic soil.
Annual Ryegrass	*Lolium multiflorum*	2 to 3 feet (60 to 90 cm)	-20°F (-29°C)	½ to 2 pounds (225 to 910 grams)	Fast growing and tolerates flooding. Absorbs excess nitrogen from soil. Can become weedy.
Winter Rye	*Secale cereale*	4 to 5 feet (122 to 152 cm)	-30°F (-34°C)	2 to 3 pounds (910 to 1,361 grams)	Deep-rooted grasses are tolerant of low fertility, acidic soils. Roots may grow back after slashing.
Winter Wheat	*Triticum aestivum*	5 feet (152 cm)	-40°F	2½ pounds (1,135 grams)	Will not die off over winter. Good nitrogen to carbon ratio.
Buckwheat	*Fagopyrum esculentum*	1 to 3 feet (30 to 90 cm)	32°F (0°C)	3 pounds (1,361 grams)	Fast-growing warm-season crop. Good for raised beds or small spaces as it breaks down quickly.
Mustard	*Brassica* species	1 to 3 feet (30 to 90 cm)	0°F (-18°C)	1 pound (455 grams)	Strong taproot mines minerals but can become a pest. Attracts bees.
Phacelia	*P. tanacetifolia*	2 to 3 feet (60 to 90 cm)	20°F (-6.6°C)	¼ pound (115 grams)	Fast-growing. Absorbs excess nitrogen and calcium in soils. Attracts bees.
Oilseed Radish	*Raphanus sativus*	2 to 3 feet (60 to 90 cm)	20°F (-6.6°C)	2 pounds (910 grams)	Grows fast with a strong taproot. Kills nematodes when tilled into soil but it may harbor Brassica family diseases.

How to Mulch

Take a page from the forest and protect the soil and plants in the garden with a healthy layer of mulch. Mulches are incredibly popular with permaculturists and gardeners because of the regenerative properties they provide for the soil. The protective layer of organic material keeps the soil temperatures more consistent, deters surface evaporation, and decomposes slowly to add nutrients. There are many things that you can use as mulch in your garden that come from nature. Avoid unnatural mulches such as plastic, shredded tires, or dyed bark, which can leech contaminants into your garden. Instead, here are thirteen natural materials that make wonderful mulches.

COMPOST

Properly finished compost is the most natural mulch in the garden. It looks tidy and clean, refreshing garden beds by framing plants with fresh, dark soil. Add 2 inches (5 cm) of compost to the soil in spring before the growing season ramps up. Spring rains will help water the microorganisms into the soil below.

LEAF MOLD

Leaf mold is partially decomposed shredded or chopped leaves. This highly nutritious organic matter breaks down quickly to make a great mulch for vegetable gardens. Pile up dried leaves in fall that are free from disease or pests and allow them to begin composting. After a few weeks or so, use shears or a lawn mower to chop the leaves into smaller pieces and apply to the garden for winter protection. In spring, topdress the leaf mold with fresh compost.

WOOD CHIPS

Wood chips are a super, all-purpose mulch for ornamental gardens containing shrubs, trees, and perennials. The attractiveness of wood chips adds a refined look to garden beds. Wood mulch also feeds beneficial fungi. Soil that is not often disturbed can create a strong mycelium layer as the wood chips decompose, just like in the forest. Wood chip mulch also lasts the longest, and it is often free or very inexpensive. Look for wood chips from arborists or local sources as they are inexpensive and readily available. Apply 1 to 2 inches (2.5 to 5 cm) of wood mulch annually as needed.

ROCKS

Decorative rocks such as riverstones, gravel, or volcanic rock do not decompose like other mulches do. Rocks do not add beneficial microbes or bacteria, but they do help reduce evaporation of soil moisture. Rocks are often used as mulch in Xeriscaping and look very attractive in desert and river gardens. Keep in mind that rocks absorb and hold both heat and cold, amplifying changes in soil temperature. Apply four sheets of damp newspaper or a layer of cardboard to the soil around plants before topping with rocks. The paper does not smother the soil like landscape fabric does, and it will eventually break down and nourish the soil.

PINE STRAW/CONIFER NEEDLES

Woodland gardens look natural when they're mulched with the needles and cones of the surrounding trees. Allow cones and needles to remain where they fall. If you need more

SOIL

mulch to spread around woodland perennials, collect fallen conifer needles from underneath similar trees and spread evenly around lower-growing shrubs and plants.

STRAW

Straw is often used in rural homesteads given the accessibility and affordability of this material. But watch out for hay, which can contain weed seeds. Use straw to create a thick layer of winter protection over annual crops, tender perennials, trees, and shrubs. In spring, remove the straw and then mulch with new straw and compost. Add a 2-inch (5-cm) layer of compost on top of beds that have longer-growing-season plants in them.

CARDBOARD

Cardboard is popular mulch because it's an inexpensive and readily available carbon source. Use cardboard as winter protection of garden beds in rainy areas to prevent nutrients leaching from the soil. In spring, remove the sheets of cardboard and compost. To mulch with cardboard at other times of the year, larger sheets should be broken into smaller 6- to 12-inch (15- to 30-cm) pieces in order to allow the soil surface to maintain enough airflow.

GRASS CLIPPINGS

We don't often think about needing to mulch our lawns, but leaving grass clippings after mowing helps create a soil-protecting mulch that slowly breaks down and feeds the grass roots. Additionally, if you grow grass long enough that the grass flowers and set seeds before you mow, mulching with grass clippings also helps reseed your lawn.

Mulching around vegetable garden plants with straw reduces weeds and evaporation from the bare soil around young plants.

LIVING MULCH

Planting low-growing groundcovers under and between plants creates a living mulch layer in your garden. Groundcovers look wonderful in ornamental gardens with perennials, trees, and shrubs that aren't often disrupted. Plant low-growing perennial and native plants that reach no higher than 12 inches (30 cm) tall and are hardy enough to be lightly tread upon so you still have access to the plants in the garden. Fast-growing groundcovers will grow happily and quickly fill in the understory.

Wild plants and volunteers make great chop-and-drop mulch. Cut them before they go to seed and avoid mulching with plants that grow from roots or rhizomes.

CHOP-AND-DROP

Chop-and-drop mulching refers to cutting back plant material and leaving it on the soil surface to decompose in place. This works for all plant material that is disease- and pest-free and allows the leaves, stems, and flower heads to decompose, which will feed the roots of their own kind, mimicking how they fertilize the soil in nature. Avoid plants that have gone to seed unless you are intending for those seeds to grow. In fall, cover chop-and-drop mulch with leaf mold and a layer of compost to build rich soil around plants with no waste.

SNOW

Snow is not often considered mulch; however, it's useful to protect the soil temperature and plant roots in cold climates. There's no need to truck in snow; just allow it to fall and sit on garden beds to insulate the soil below and protect it from dips in temperature.

AQUATIC PLANTS

If you have a pond or other water feature you may have aquatic plants that can be vigorous or—even invasive—in a water garden environment. Thinning aquatic plants allows you to use this material to mulch dry land gardens without risk of spreading invasive species because the conditions are too dry for these plants to do anything but decompose and protect the soil.

SEAWEED

Shredded seaweed works well as mulch in seaside gardens where plants can naturally handle salt spray and salt accumulation in the soil. Be aware that too much salt can cause injury to plants, so this is only appropriate in coastal gardens where salt is tolerated. Collecting seaweed should be done sustainably, only foraging for what you need and removing seaweed from the beach rather than live seaweed from the ocean. Seaweed used as mulch should be chopped or run over with a lawn mower prior to applying it to the garden. Seaweed mulch is often best when mixed with another mulch such as leaves or chop-and-drop.

Sheet Mulching

Plant directly in the compost layer in the top of a newly sheet-mulched bed. Mulch around new plants to protect the soil and speed up composting.

- Lay down a paper layer at soil level and top it with mulch.

- Collect materials from your property or locally to recycle waste products.

- Build soil with alternating carbon and nitrogen layers, and then top with manure, compost, and a final layer of mulch.

Sheet mulching, sometimes referred to as sheet composting or lasagna gardening, is the process of building a new garden bed of soil in layers, or sheets, of various carbon and nitrogen source materials. Building a garden bed this way is a turnkey method for reclaiming land as garden beds to build soil, suppress weeds, and mulch all in one.

Sheet mulching can be as simple as applying a layer of newspaper or cardboard on top of weedy soil and covering that with 12 inches (30 cm) of mulch material that will decompose quickly. Composting in places like this is very fast, so you can reclaim that weedy soil in fall to be ready for spring planting. It allows the soil microorganisms to become active and thrive

without threat of tilling that will disturb them.

In my experience of creating new soil with sheet mulching, it can be pretty successful using just about any organic material you have on hand, such as cardboard or leaves. That being said, layering a combination of carbon and nitrogen materials helps speed the process and build a deeper soil bed. From the ground up, here are some layers you can choose to add:

- Moistened paper (four sheets thick) or cardboard to cover the area where you want to suppress weeds. (Remove any glossy pages from the newspaper, as well as plastic, staples, and tape). Other organic materials such as fabric can also be used but they will take longer to decompose.

- 12 inches (30 cm) organic materials as listed in the sidebar on the next page

- 1 inch (2.5 cm) manure

- 2 inches (5 cm) compost

- 1 to 2 inches (2.5 to 5 cm) of a top mulch layer such as wood chips, leaf mold, or straw

MAKE IT!

1 Remove any plants that have gone to seed. Mow or cut the remaining weedy vegetation and leave the clippings in place.

2 Lay down cardboard or sheets of paper and wet each layer as you set it in place.

3 Balance the organic materials as you would compost using a combination of nitrogen and carbon sources. A layer that is too high in nitrogen (such as green garden clippings) will quickly become wet and slimy while a layer that is more carbon heavy (such as bark mulch) will take longer to decompose and will not add enough nitrogen to the soil.

4 Next, the manure layer is to add nutrients to the base of the planting bed while the lower layers decompose.

5 The top layer of compost is the planting bed for seeds and plants.

6 The final mulch layer is for moisture retention and weed suppression on the top layer. Pull back or remove this layer in spring for planting.

ORGANIC MATERIALS TO USE IN SHEET COMPOSTING

Nitrogen Sources
- Animal bedding with manure
- Coffee grounds
- Composted manure
- Garden/grass clippings
- Kitchen scraps
- Seaweed

Carbon Sources
- Animal bedding
- Autumn leaves
- Finely ground bark
- Grain hulls
- Sawdust
- Straw
- Wood shavings

SOIL

Hugelkultur

- Mound slow-to-decompose materials such as rotting logs and branches, cover with a carbon layer, and then add a top layer of compost that can be planted.

- Add other slow-to-decompose materials such as fruit pits and eggshells to the mound. Add more layers on top of the compost such as manure and carbon sources, and then top with more compost and mulch around the plants.

- Plant hugelkultur with biodynamic or wild plants.

A hugelkultur is a mash-up of a slow/cool compost pile and a garden bed. It allows you to create a foundation using materials that take a long time to compost at the bottom of the bed, such as rotting logs, branches, sticks, woody vines, eggshells, and fruit pits. This woody layer is covered with carbon-rich materials and compost. Green waste and manure can also be added, similar to the layers in sheet mulching. The mound is finished with a layer of compost that can planted. The materials in the center of the beds take a long time to break down, so this compost pile is useful as a self-feeding, long-term garden bed. It is beneficial to plant biodynamic or wild plants, which will continue to act as soil fixers, break down the organic matter, and add nutrients to the soil.

A hugelkultur is well-suited to a woodland or forest in order to efficiently use fallen tree branches and create rich beneficial garden beds and soil. This concept can also be created in home gardens as a feature for growing all sorts of annuals, perennials, food crops, and ornamentals. Hugelkultur gardens are very good at holding water and heat as they decompose, so plants will enjoy early crops and longer seasons.

The word *hugelkultur* means "hill mound," which can be seen in the structure as it resembles a miniature hill. The basin is filled with slow-to-decompose materials such as rotting wood and then layered with carbon and nitrogen waste and topped with compost.

This new hugelkultur was planted with sage and mullein to increase soil fertility.

MAKE IT!

1 Map out the location that you would like to build a hügelkultur using flour or string.

Dig the hole at least 18 to 24 inches (45 to 60 cm) deep.

2 Add rotting logs, sticks, and twigs to fill the basin and create a mound.

3 Fill in the spaces between the logs and branches with other slow-to-decompose materials such as woody vines, eggshells, and fruit pits.

4 Add a layer of carbon material (such as paper, straw, or leaves) over the top of the woody layer.

5 The top layer can be 6 to 8 inches (15 to 20 cm) of compost, which you can plant right into. However, feel free to add more layers to the mound. A layer of green waste on top of the compost will speed decomposition. A layer of well-composted manure will feed the plants while the layers below decompose. If you add these additional layers, then top with a final 6 to 12 inches (15 to 30 cm) of compost on top for planting.

DECOMPOSING WOOD AS WATER RETENTION

While most compost and mulch systems require the organic material to be in small or chopped up pieces in order to accelerate decomposition, there is huge value in adding large pieces of rotting wood and branches to gardens to hold moisture and encourage beneficial fungi. In new garden beds, this helps to inexpensively fill the space without a need to bring in heaps of new soil while at the same time increasing water-holding ability. The space around the wood is then filled with compost, organic matter, and carbon materials to create a slow compost pile. Top the bed with a foot of compost and plant directly into it. This type of bed will need much less watering throughout the season and works well for nutrient-hungry crops such as blueberries or potatoes.

With raised or in-ground garden beds, large pieces of rotting wood or branches can be used to fill the base of the bed and act as a sponge to hold water.

Wild Plants and What They Say about Your Soil

> - Allow weeds to grow in wild spaces and parts of the garden.
> - "Read the weeds" to learn about soil needs.
> - Cultivate wild plants as soil fixers.

Think about what happens when we clear out plants from wild spaces and leave the soil bare. The soil dries out or becomes otherwise lifeless *dirt*. And what does nature do? She sends in hardy wild plants that thrive in poor soil to help bring it back to life. While some may call these wild plants "weeds," perhaps we should refer to them as "soil fixers." These plants have arrived and are thriving in your garden because the soil has summoned them. We would benefit by reading the messages that they are providing.

Let's speak "plant" for a moment. Plants give us a whole lot of information about the world around us. Wild animals know how to read these messages. They know which plants to eat, which to use as medicine, and which to use as building materials. By contrast humans have taken a different approach by labeling plants as "good" or "bad" and then altering the landscape to cultivate only the ones we want to use. While animals are learning from plants, we are trying to dominate them.

Of course, our first thought is to remove the weeds so they don't compete with our cultivated varieties, yet leaving them to grow provides a host of benefits. The roots dig deep and loosen compacted soil. The leaves and roots both add organic matter to the soil, from above and below the surface. As they decompose, they add essential nutrients back into the soil. It seems to me that we should be thanking these wild plants for their service, as opposed to cursing them for reappearing whenever we rip them out.

"When humans make a clearing, nature leaps in, working furiously to rebuild an intact humus and fungal layer, harvest energy, and reconstruct all the cycles and connections that have been severed. A thicket of fast growing pioneer plants, packing a lot of biomass into a small space, is an effective way to do this."

—Toby Hemenway, author of *Gaia's Garden*

Red clover and dock/common sorrel work together to improve poor soil. Clover fixes nitrogen and dock's taproots break up hard soil.

SOIL FIXERS: WILD PLANTS AND HOW THEY HELP SOIL [9]

Common Name	Botanical Name	Indicates	Hardy to
Bindweed / Morning Glory	*Convolvulus arvensis*	Poor drainage, compacted soil	Breaks up hardpan soil, adds minerals to soil through root decomposition.
Chickweed	*Stellaria media*	Nitrogen-rich soil, iron accumulation	The nutrients magnesium, potassium, phosphorus, and manganese are released into the soil when it decomposes. Edible source of vitamins C and B, plus minerals.
Clover	*Trifolium repens*	Low nitrogen, heavy, acidic soil	A cover crop that helps fix nitrogen and feeds insects. Accumulates potassium and phosphorus.
Dandelion	*Taraxacum officinale*	Poor drainage, acidic, compacted soil but also fertile, well-drained soil	Fixes heavily compacted soils as its taproot breaks up soil. Taproots mine for calcium, iron, and other minerals from deep in the soil, bringing them up to soil level. Helps feed insects and all parts (leaves, flower, and roots) are edible.
Horsetail	*Equisetum arvense*	Acidic, light, sandy soil with good drainage	Accumulates silicon, magnesium, calcium, iron, and cobalt, which are released into the soil when it decomposes.
Mullein	*Verbascum thapsus*	Dry, crusty, or compacted soil with low fertility	Accumulates magnesium, sulphur, and potassium.
Plantain	*Plantago major*	Heavy, compacted, acidic soil with poor fertility	Deacidifies soil. Adds calcium, magnesium, silicon, sulphur, manganese, and iron when it decomposes.
Quackgrass	*Agropyron repens*	Poor drainage, clay soil.	Netlike root system helps with erosion. A natural preventative for slugs.
Scotch Broom	*Cytisus scoparius*	Low fertility, acidic soil	Fixes nitrogen.
Sorrel	*Rumex acetosa*	Acidic soil	Alkalinizes soil. Mines for calcium, phosphorus, and minerals that can be turned under to balance pH.
Stinging Nettle	*Urtica dioica*	Acidic, heavily cultivated, compacted clay soil	Helps break up heavily compacted clay and accumulates nitrogen, copper, potassium, and sulphur. Edible and medicinal plant that is highly nutritious.
Thistle	*Cirsium*	Heavy, compacted soil	Its taproot breaks up soil and brings up iron and moisture for shallow roots. Taproots can penetrate as deep as 20 feet (6 meters) into the soil.
Vetch	*Vicia atropurpurea*	Low nitrogen, low fertility	Fixes nitrogen and is often used as a cover crop. Accumulates potassium, phosphorus, copper, and cobalt.
Yarrow	*Achillea millefolium*	Low fertility, low potassium, sandy soil, parched soil.	Improves fertility of poor soil, adds organic matter.
Dock/Sorrel	*Rumex* spp.	Acidic soil, poor drainage	Taproot breaks up soil. Taproots mine for iron, calcium, potassium, and phosphorus.

After clearing the wild plants from our soil, we then plant many edible and ornamental plants that can't naturalize (often because we eat them or harvest them). Our soil becomes strained, and the seeds of wild plants grow in order to help restore soil health.

2
Water
Efficiency in Collection and Use

Capture and store rainwater for garden irrigation.

Recycle water and use it as many times as possible.

Live within your local water budget—don't use more than your rainfall. Meet all garden needs with captured, cached, or recycled water.

The former project manager for the International Landed Learning Program (a school-to-farm program I used to volunteer for) is writing a children's book following the path of a single drop of water. It's a beautiful and fascinating story that shows us how each drop of water we use has both a history and a future. She shared the story with me over lunch one day, and it brought me to tears. The total amount of water that we have in this world is finite, and, sadly, much of it is becoming polluted and unusable as drinking water. Her storybook helps illustrate the importance both of protecting and respecting water because this precious resource is essential for all the organisms on our earth to survive.

As we design our regenerative gardens, water catchment, conservation, and management systems are some of the most important projects we can include in our spaces. Australia and California have faced extreme drought situations for many years, and as result some very smart water solutions have been developed. I've personally seen a dramatic change just in the twenty-five years I've lived in Vancouver, British Columbia, a city known for its rainy climate. The municipal systems that were developed to cache water are no longer large enough to provide water throughout the summer months as rainfall is lessening. In recent years, our water demand in summer has surpassed the reserves because drought conditions are paired with a high demand for water to extinguish wildfires. This goes to show that even in places known for their rainfall, collecting water, purifying it, and storing it for future use is a big part of building a home ecosystem.

As water needs and rainfall can vary dramatically depending on where you live, the first step is to measure and record how much water you can collect from rainfall and how much water you need to irrigate the garden. Note any periods of drought to calculate how much water will need to be captured to meet the garden's needs during a drought. Use these measurements to help choose which projects from this chapter you will add to your regenerative garden.

Rain Barrel

- 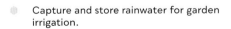 Capture and store rainwater for garden irrigation.

- 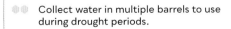 Collect water in multiple barrels to use during drought periods.

-  Meet all garden needs with collected water.

A rain barrel is a water-collection method that redirects rainwater into a storage system so that it may be used as future garden irrigation. Rain barrels can be purchased from municipalities and home-supply stores or made from food-safe drums that are plastic, metal, or wood (such as wine or whisky barrels).

Rainwater is directed through a series of gutters planted with drought-tolerant *Sedum* that help filter the water prior to returning to the ground. The top two gutters are mounted at angles to direct runoff to the planter below.

HOW SAFE IS ROOF RUNOFF FOR GARDEN PLANTS?

Often rain barrels are set up to collect rainwater from the rooftops of structures surrounding the garden. A solid, non-porous structure such as a metal roof is ideal for water collection. Tile and shake (wood shingle) roofs can also be used for water collection as long as the roofing material is untreated. The runoff from asphalt and rubber roofs should be avoided or used to irrigate non-edible crops as they can be high in anaerobic bacteria and petroleum products. A galvanized zinc roof is not recommended for garden irrigation to avoid metal toxicity (seen as stunted growth and leaf discoloration).

Keep in mind that these barrels typically have a capacity of around 50 gallons (189 liters), which can water 80 square feet (7.5 square meters). You can certainly connect a few barrels together through the overflow pipes to gain more storage. Or install a cistern, which is built just as these instructions demonstrate with a larger barrel that can hold 200 to 1,000 gallons (757 to 3,785 liters) more capacity. If you have a small garden a rain barrel can offer some supplementary water, but in periods of drought or for larger gardens, a number of rain barrels, a number of cisterns, plus a combination of the other water catchment systems presented in this chapter allow more watering needs to be met.

This rain barrel design was modeled after the design created by the publication *"Installing a Cistern"* created by the Seattle Public Utilities and King County Wastewater Treatment Division's RainWise Program, www.700milliongallons.org

SITE PREPARATION

Install a rain barrel on level land near a downspout. Prepare the ground below by compacting the soil or creating a bed of sand, gravel, or concrete. Look for a shaded location to place the rain barrel to keep the water from getting too hot.

Plan where the overflow pipe will discharge. Situate the overflow runoff 5 to 15 feet (1.5 to 4.5 meters) from buildings. At a minimum, the overflow drain should end 5 feet (1.5 meters) from structures without basements and 10 feet (3 meters) from structures with a basement height of 5 feet; add 2 additional feet (0.6 meter) for every foot (0.3 meter) deeper the basement is than 5 feet. Be sure to check with local municipalities for guidance on exact measurements required for your home.

❀ ❀ ❀ ❀ ❀

MATERIALS

- A large barrel (choose a dark color to reduce algae growth)
- 3- to 4-inch (7.5- to 10-cm) overflow pipe and connectors in ABS or PVC plastic
- A self-cleaning leaf filter with screen
- A finer screen for the barrel opening
- A ¾-inch (2-cm) pipe with faucet as the drain that connects to a garden hose
- A 3- to 4-inch (7.5- to 10-cm) plug to create a cleaning hole in the base

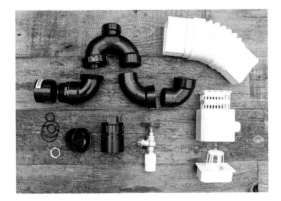

MAKE IT!

1 Create a level foundation by compacting the soil, then adding 4 to 6 inches (10 to 15 cm) of packed sand, concrete pavers, or poured concrete. Use a carpenter's level to check the base before placing the barrel on top.

2 Cover the inlet hole at the top with fine window screen.

3 Add a self-cleaning leaf filter to the downspout directly above the inlet. Connect the gutters to the leaf filter.

4 Install the overflow pipe at the top of the barrel by cutting a hole in the barrel and attaching a 3- to 4-inch (7.5- to 10-cm) pipe in the hole. Connect a p-trap at the top of the overflow pipe to prevent mosquitoes and rodents from getting in.

5 Route the overflow pipe away from the structure.

6 Add a drain valve by cutting a hole in the barrel to insert a faucet that connects to a garden hose.

7 Cut a cleanout hole in the bottom of the barrel and seal with a removable plug. The hole should be large enough to spray a hose inside the barrel and insert a long-handled brush to scrub the bottom.

In rainy climates, leave the faucet partly open to slowly drain as rainwater is collected. In spring, close the valve to collect water for summer use. Be sure to run the hose away from any structures when the barrel is draining during the rainy season.

Olla Water Catchment System

- Bury terracotta pots in the garden and cover each with a glazed saucer as a lid.

- Build a flute-shaped olla by sealing two terracotta pots together and burying it in the garden.

- Set up rainwater catchment or greywater recycling system directed toward the ollas.

An olla is a water catchment system made from unglazed terracotta pots that are buried in the ground and then filled with water. The unglazed clay then slowly irrigates the soil as water is only being pulled from the pots when the surrounding soil and plant roots reach for it. This efficient slow irrigation system is thought to have originated in Northern Africa and China more than 4,000 years ago. Today ollas are used around the world to helps to regulate irrigation and provide deep root watering to crops planted adjacent to the pots.

It's fairly easy to make an olla catchment system for the home garden. In its simplest form, a terracotta pot with no drainage hole is buried in the ground and covered by a glazed plant saucer. A more advanced design is to create a "flute" that has a smaller surface area above the soil line. This reduces evaporation and goes deeper into the soil.

Create a fluted olla using two unglazed terracotta pots, a clay saucer, and some water-proof silicone sealer. The wide openings of the two pots are sealed together and one of the drainage holes is filled to create a watertight

The glazed plant saucer is placed on top to prevent insects, such as mosquitoes, from entering the olla and breeding. The top saucer can be glazed in order to reduce evaporation above the soil. This ensures that all the water will seep out through the clay in the soil. Unglazed saucers are also effective and will greatly limit evaporation and have the same effectiveness at preventing mosquitoes from breeding.

Unglazed clay pots cannot withstand a winter freeze and therefore need to be removed before winter and washed and stored until the next gardening season.

vessel with a small opening on top (the other drainage hole). The olla is then buried into the ground with just the neck above the soil to allow filling with water.

Ollas are "planted" in garden beds, spaced between plants so that they can provide supplementary water and reduce the time between garden watering. Ollas are also very effective in garden pots to reduce watering during the thirstiest times of the year. For example, an olla planted in a 20-inch (51-cm) container can reduce watering from every day to every four to five days in the heat of summer. Ollas are particularly valuable with new plants, seedlings, and seed starting to help young plants get established in the garden.

WHAT SIZE OLLA DO I NEED? [10]

For large (20- to 24-inch) (50- to 60-cm) containers, plant a 6- to 8-inch (15- to 20-cm) olla. For garden beds, bury multiple 8-to 12-inch (20- to 30-cm) ollas.

◎ ◎ ◎ ◎ ◎ ◎

MATERIALS
- 2 unglazed terracotta flowerpots
- 1 glazed terracotta saucer
- 1 coin slightly larger than the drainage hole
- Waterproof, food-grade, 100% silicone caulk

MAKE IT!

1 Use a coin that is slightly larger than the drainage hole to cover the hole in one of the pots. Run a thin line of silicone sealer around the hole and attach the coin. When it's dry, add another ring of sealer around the seam and smooth with your finger. Turn the pot over and fill the hole with silicone, smoothing the seam with your finger again. Allow the silicone to dry completely. Test the seal by filling the pot with water and looking for any leaks.

2 Run a line of caulk around the rims of both pots. Place one onto the other. Smooth the silicone seam with your finger or a sponge. Allow the sealer to dry according to the instructions on the package of caulk.

3 When the silicone is dry, run another bead of caulk along the seam between the pots. Use your finger or a sponge to smooth the caulk around the seam for a tight seal. Allow to dry according to the instructions on the caulk.

4 Test the olla by filling it with water. Log how long it takes for the water to seep through its sides.

5 Dig a hole where you plan to plant the olla and submerge.

6 Backfill with soil, leaving only 1 to 2 inches (2.5 to 5 cm) of clay above the soil, fill with water, and place the glazed plant saucer on top.

Water should be added to an olla when the water level inside falls below 50 percent to ensure that the porosity of the clay remains intact.

This olla is filled by a solar irrigation system that is attached to a rain barrel. On sunny days, the solar panel encourages the pump to draw more water from the barrel than it does on rainy days.

WATER FOR DEEP ROOTS

One of the best ways to reduce the need to add supplementary water to the garden is to teach plants to reach down deeply and find it themselves. Think about trees planted in an urban environment such as a city sidewalk or public park. The roots of the trees are conditioned to find water that has been stored in the soil from extended periods of rain. As long as the soil is full of organic material to hold onto that moisture, it can remain there until the plants need it. In desert climates and where drought is prevalent, it's even more important to add plenty of organic matter to the soil to hold onto what little water there is naturally.

Watering *deeply* and *infrequently* trains plant roots to grow down into the soil and get better at accessing water stored deeply. This creates sturdier plants that need less supplemental water.

DIY RAIN GAUGE

Making a rain gauge is as simple as putting an opened glass jar or plastic container outside in your garden and collecting rain. Measure rainfall by inserting a wooden ruler or stick into the water, straight down and touching the bottom of the container, and then noting the measurement in inches or centimeters of the part of the wood stick that is wet. This can also be used to measure sprinkler water. Create a one-step solution by drawing measurements on a Mason jar or clear plastic container with a permanent marker or marking clear tape with the measurements and attaching it to the jar.

The Darwin Australia Living Smart Project estimates that, "Generally, most plants need about 30 mm of water each week to be healthy. Natives often need less, though, more like 30 mm every two weeks." [11]

How much rainfall your garden needs can depend on many factors: how fast it drains, the amount of organic material in the soil that will hold moisture, and how much surface evaporation there is from heat or lack of mulch. The types of plants that are in the soil are also important, as drought-tolerant plants, established perennials, trees, and shrubs need less supplemental water while annuals with shallow roots need more regular irrigation.

Determine how much rainfall your garden needs by sticking your finger into the soil up to your first knuckle after a rain. If the soil at the tip of your finger feels cool and sticks to your skin when you remove it, the soil is wet. Is your finger comes out dry and the soil feels warm, the garden will need supplemental water.

SOLAR DRIP IRRIGATION

When it's sunny out you'll need to water the most. That's why a solar-powered automatic watering system is so brilliant! Solar drip watering kits are powered by the sun to water more often when it's sunny out and less often when it's not. They include a solar battery that regulates the water flow and frequency. Hook this baby up to a rain barrel and your garden watering is completely hands-free.

Build a Rain Garden

While a rain garden sounds fancy, it's really quite a low-maintenance system used to filter and release stormwater runoff, keeping water local to the soil and creeks as it was intended by nature.

- Disconnect downspouts from sewers and allow rainwater to return to the earth.

- Redirect rainwater into a rain garden.

- Choose wildlife-attracting and native plants for the rain garden.

Home gutters are sometimes directed into the sewer systems, which route stormwater through the same treatment system as toilets, showers, and sinks. Not only is this treatment unnecessary, it can also cause overflows of the sewer system that redirects into the nearest river, lake, or ocean. This leaves local soils and waterways devoid of the rainfall they need to stay healthy.

At minimum, disconnecting downspouts and redirecting them into the surrounding garden allows for the runoff to go into a rain barrel or cistern, pond, lawn, garden, or a rain garden. This is a more efficient use of the water and allows it to be used as irrigation while it naturally filters through the earth.

The basin of a rain garden is filled with plants that love moisture and can act as biofilters to purify the water. The upper swales have drought-tolerant plants that love to grow deep roots and access the available groundwater when they can. The design can be as simple as attaching a gravel-filled trough to a downspout and building a garden bed around it to designing more elaborate rain gardens that become a haven for bird and insect species.

SITE PREPARATION [12]

Locate a rain garden 5 to 15 feet (1.5 to 4.5 meters) away from buildings. At a minimum, the overflow drain should end 5 feet (1.5 meters) from structures without basements and 10 feet (3 meters) from structures with a basement height of 5 feet (1.5 meters); add 2 additional feet (0.6 meter) for every foot (0.3 meter) deeper the basement is. Don't locate a rain garden over underground utilities or large tree roots.

The rain garden's location should be sloped away from buildings to direct overflow out through spillways to other gardens rather than running back toward structures; however, the garden itself should have no more than a 5 percent grade overall (1 foot drop in 20 feet [30 cm in 6 meters]).

In rainy climates, the bottom inside of the garden should be 15 percent of the square footage of the runoff source. So a 500-square-foot (46-square-meter) roof requires a 75-square-foot (7-square-meter) garden base (500x0.15=75) (46 x 0.15 = 7).

Often plants will move into the right position in the garden, like these drought-tolerant plants that have replanted themselves to the basin of this rain garden.

- The garden is built as a swale: a recessed center about 4 to 8 inches (10 to 20 cm), with berms around the perimeter to hold water in. The center planting area should be level to prevent pooling.

- The garden is generally twice as wide as it is long, with the widest part at the lowest point of the slope.

- Choose a location that can handle plenty of water saturation, away from septic systems and plants, shrubs, or trees that don't like their roots to stay wet.

- The rain garden design will be planted with water-loving plants and could have spillways to other garden beds that the overflow can irrigate.

Water the plants regularly to establish them in the first few years. Add more compost as mulch annually.

MAKE IT!

1 Determine the garden's location and extend the downspout to the highest point of the garden.

2 Use a garden hose or sprinkle flour to outline the garden's shape.

3 Dig the base 24 inches (60 cm) deep and use the soil to berm up the sides.

4 Fill the base with 12 inches (30 cm) of a rain garden soil mix (one-third compost and two-thirds garden soil). This leaves 12 inches (30 cm) of ponding depth (the space from the top of the base soil to ground level where water can collect during rainy periods).

5 At the lowest point of the garden edge, create an overflow area that is packed with rocks. This allows the overflow to release without eroding the garden. Overflow should be directed to a street drain.

6 Choose the right plants for your garden's conditions. Suggested plants can be native plants or cultivated garden favorites that perform well in your unique microclimate. The best place to find the right plants for your garden is at your local garden nursery, which will carry both native and cultivated plants.

7 Plant the garden plants, mulch with 2 inches (5 cm) of compost, and water well.

Build a Wicking Bed

- Create a water reservoir under the soil of a garden bed.

- Use recycled materials to build the bed.

- Capture rainwater runoff to fill the reservoir.

A wicking bed is like a giant self-watering container in your garden. Self-watering containers are quite popular because they capture water runoff and store it in a reservoir that is available for the soil and plant roots as needed. This system can exponentially, reduce watering and it also creates strong roots and sturdy plants.

Wicking beds can be built in the ground or as raised beds. The base of the reservoir is

A wicking bed will look just like a regular garden bed; however, below the surface there's a reservoir to hold water that will dramatically reduce the water needed throughout the growing season because there is less evaporation when water comes from below the surface.

created with an impermeable barrier below to hold water, drainage pipes for water flow, and gravel to store the water and create a base for the soil. Above the gravel is a permeable divider like landscape fabric, and above that the soil and plants.

Wicking beds can be filled by overflow spouts directed to them, through rainfall, or manually. Multiple beds can be connected together to fill one another through the overflow spouts. In summer, the reservoir will need a weekly top up. In spring and fall, the bed should only need to be filled every 2 to 3 weeks, or less if you are in a rainy climate.

To prevent mosquitoes from breeding in the reservoir, cover the opening of the reservoir whenever it is not being filled. Cedar fence posts are used for this project because they will bow less than standard lumber over time. The raised bed frame can be made with any material that is on hand, such as cedar planks or brick. Avoid treated materials to reduce soil contamination.

❋ ❋ ❋ ❋ ❋ ❋

MATERIALS

- 12 cedar fence posts cut the length of the garden bed
- 3/8- x 10-inch (1- x 25-cm) galvanized steel spike nails
- Pond liner measured to the inside dimensions of the bed plus 12 inches (30 cm) of additional width on all sides
- Staple gun and staples
- Weeping tile (a porous drainage pipe)
- 3/4-inch (1-cm) gravel; use "clear" gravel, which does not have smaller pieces of gravel that will fill up the spaces in between
- Dishwasher drain tube to create a fill tube
- Landscape fabric to prevent soil from filling up the piping and the spaces between gravel

This project was created and photographed by author and horticulturist Steven Biggs and his children in their Toronto, Ontario, home. Steven's teenaged daughter, Emma Biggs, is also an author and avid tomato grower. They built these beds for Emma's tomatoes to prevent water stress and to permit Emma to grow tomatoes near a black walnut tree which would otherwise poison them. Steven and Emma can be found at www.foodgardenlife.com.

MAKE IT!

1 Prepare the site by removing weeds, compacting the soil, and leveling the ground.

2 Cut the posts to length (if needed) and notch the ends by cutting half the depth of each post as deep as the width of the connecting post. Place the two notched side posts directly on the ground, notch side up.

3 Set two end posts across the ends of the beds, with the notches facing down on the notches of the side posts. The four base posts should sit flush to the ground and create a square or rectangular bed.

4 Nail spikes into the corners of the posts to keep them in place and add subsequent layers, nailing them in place until the bed reaches the desired height. In this case, they are three posts high.

5 Install the liner at the bottom by placing it on the ground and up about 8 to 10 inches (20 to 25.5 cm) at the side. Secure temporarily with staples to keep it in place until the gravel pins it into place.

6 Place coils of weeping tile in the bottom. The tile permits water to quickly move through the reservoir, and it also holds up the soil above.

7 Connect a fill tube (a piece of drain hose installed into the weeping tile) to the weeping tile. The fill tube allows the reservoir to be filled with a hose after soil has been added.

8 Add the gravel. It supports the weight of the soil above it while the spaces between the pieces of gravel fill with water. Water moves upwards through the gravel by capillary action.

9 Cover the gravel with landscape fabric to keep soil out of the reservoir area. Create a depression in the one corner. While in theory water wicks up the gravel, this soil-filled wick dips into the reservoir to ensure that water can wick up.

10 Add 12 inches (30 cm) of soil to the top of the bed and plant directly in it.

Self-watering pots are a great way to reduce watering container plants. However, if self-watering pots are used improperly, they can also cause problems by providing a home for breeding mosquitoes, fungus, and root disease.

- Create reservoirs in pots to reduce watering.
- Use large pots to create mini ecosystems.
- Use recycled water to fill the reservoirs.

A self-watering pot is designed with a reservoir to hold excess water below the soil. Plant roots are able to reach below to wick up water into the soil as it becomes dry. Some self-watering pots are double-walled and hold the reservoir of water between the walls. This adds extra insulation to the planter and means the container can hold a whole lot of water.

Some plants need excellent drainage and will become stressed if the soil is consistently wet. Plants such as succulents could dislike the moisture in a self-watering pot and would be much happier if the soil were allowed to dry out between watering. Research a plant's individual needs before planting it and choose the right plants for self-watering pots.

PREVENT A BREEDING GROUND FOR MOSQUITOES

The drainage holes on some pots are sneaky places for mosquitoes to reproduce. They can lay eggs in just a teaspoon of water so the reservoir is an ideal breeding ground. To prevent this from happening, flush the water every few days to a week or plug the hole with a metal or plastic pot scrubber so they can't fly in, but water can still get out.

MANAGING MOISTURE LEVELS IN THE SOIL

Self-watering pots are best used to extend the time between watering as opposed to keeping the soil constantly wet. Dry soil will draw up moisture and hold it so that it never dries out. This system can be too efficient and cause soil to be oversaturated. It's best to let the reservoir dry out so that the soil drains between watering.

WATER

55

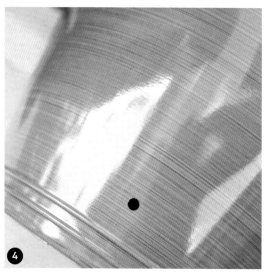

MATERIALS

- Two similarly sized garden pots that are stackable; at least one of the pots will have no drainage holes
- Electric drill
- Wood blocks or a brick as risers (may not be needed)
- Thick cotton string

MAKE IT!

1 Acquire two pots to make a self-watering planter. The pots can be sealed ceramic or plastic and can be garden pots or other containers such as storage bins or 5-gallon (19-liter) buckets.

2 One pot should have no drainage holes; the other needs drainage holes, which can be added by drilling holes in the bottom of the container.

3 Stack the pots so that the inside pot (the one with drainage holes) sits a few inches above the bottom of the outside pot (the one without drainage holes) to create a water reservoir. If the pots are too close together, use a wire trivet, a brick, or some wood blocks as risers to elevate the inner pot.

4 For outdoor pots, create an overflow hole in the outer pot just below the level where the inner pot's bottom sits. This allows excess water to escape so that the soil doesn't become soggy.

5 For smaller pots and indoor pots, simply check the pot after watering and empty the excess water so that the reservoir is no more than three-quarters full.

6 There is no need to add cotton string for wicking up the water in plants with deep roots; the plant's roots will grow down and draw up the water. With shorter-rooted plants, though, add cotton string as a wick into the soil by cutting a length of string that will sit in the reservoir and 2 inches (5 cm) into the soil. Tie a knot 2 inches (5 cm) from the top of the string and thread it through one of the drainage holes so that the longer end of unknotted string sits in the reservoir.

7 Add potting soil mix and plant as you would normally pot up a container.

Deep-Watering Tube
for Container Gardens

Container gardens that are deep, such as strawberry and herb pots, can benefit from the addition of a slow-release irrigation tube to slow the water and redirect it more evenly.

- Create a watering tube to more efficiently reach plant roots.

- Use recycled materials.

- Fill watering tube with captured rainwater.

A strawberry jar may be a smart solution to keep ripening fruit off the ground but there is no way to properly deep-water the plants. Even using a watering can with a narrow spout to direct water into each of the pockets will still not saturate the soil in the center of the pot. You are doomed to have shallow roots and thirsty strawberry or herb plants. This next simple solution will direct water down to where it is needed and help you grow more resilient strawberry and herb plants.

MATERIALS

- Pocket-style strawberry jar planter
- Tape measure
- 1¼-inch (3.1-cm) plumbing pipe
- Pipe cap for ¼-inch (6 mm) pipe (shown in orange)
- Pipe lid for ¼-inch (6 mm) pipe (shown in black)
- Handsaw
- Electric drill

MAKE IT!

1 Use a tape measure to determine the inside height of the strawberry pot. Measure from the bottom of the inside of the pot to the top lip of the pot. The cap will sit just slightly above the soil line. Cut the pipe to length with a handsaw and then drill holes along the length and diameter of the tube with a ¼-inch (.6-cm) drill bit.

2 Attach the orange cap to the bottom end of the tube and set it in the center of the pot. Put the black lid on to prevent soil from getting in the tube while you fill the planter. Fill the planter with soil, holding the tube in place as you go.

3 Add the strawberry plants to each of the holes and two or three at the top around the watering tube.

4 After planting, thoroughly soak the entire jar to fill the soil in around the roots of the new plants. Don't let the soil dry out for a week as they adjust.

To water, lift the lid and pour water into the tube and let it flow into the soil until you can feel moisture under the plants.

3
Plants
Growing Life

Include plants in your garden that are right for your mini ecosystem.

Test, observe, and listen to plants.

Include native and local plants.

It's hard to imagine where we would be without plants. Plants provide us with our food, medicine, building materials, firewood, transportation, animal food, dye, and ceremonial effects. In our home gardens, we have traditionally chosen plants based on what we've adopted in our gardens, what friends and neighbors give to us, the food we want to grow, and (let's be honest) what looks pretty at the garden center. More and more home gardeners are evolving in the way that they think about plants in their garden. In addition to an attractive and productive space, they see the garden as a small landing pad for wildlife and part of the local ecosystem. It can also be a community space to brighten the days of passersby, share plants and harvests, and delight children. You will see more of these types of projects in the community chapter, but in this chapter, we're going to focus on the plants we choose for a regenerative garden.

There's no doubt you've heard the common gardener's mantra, "Right plant for the right place." Choosing plants that work within your ecosystem and climate naturally will give you the highest chances of success. But there is so much to learn about how plants work together that we are just beginning to understand. It's not just soil pH, temperature, rainfall, and light requirements that define what is "right." There are many more components to plants that are needed to work together within the environment. When we're adding plants to our gardens, some may seem to be the right ones on paper but they may not perform as expected. Or we might have great luck in one area of the garden with a plant and not in another area. It's all about testing, observing, and listening to what the plants are trying to tell us.

Of course, one of the easiest ways to do this is to learn about our native plants and include them in our landscape. This is not to say that an entire garden needs to be planted with native plants. I've found this subject to be a common debate as we continue to grow as ethical and responsible gardeners. Remember at the beginning of this book I described the principle of "good, better, and best"? It's an example of the permaculture Transition Ethic that allows us imperfection while we continue to educate and grow our commitment to regenerating the ecology around us. If you can include native plants in your garden along with the plants you have adopted, those that have been given to you by friends and neighbors, the food you want to eat, *and* what looks pretty at the garden center, you are doing so much to create a garden that serves both you and the land in the "right" way.

CHOOSING PLANTS
FOR YOUR GARDEN

This book is intended to be an idea guide
that sparks your interest in creating a regenerative garden, yet it would be impossible to list the
exact cultivars that are right for your unique garden.
For many of the projects, I have included a "honeycomb"
plant list with examples of plants that can help get your started.
Take the ideas and plant lists to your local garden nursery to find
the right plants for your specific area. Some native plants
can be classified as invasive in other regions, and, of
course,not all plants thrive in all climates. By noting
some of the plants that are often well-suited
to the projects, you should easily be able
to tailor the plant choices that best suit
your unique garden.

Interplanting Herbs

Cabbages and other Brassicas are susceptible to many pests, so plant them in between a variety of herbs for both the aesthetics and protection.

- Plant herbs throughout the garden.

- Plant companion plants to deter specific pests.

- Plant a wide variety of plants throughout the garden to increase diversity and resilience.

The idea of planting fragrant herbs and vegetables as companion plants in the garden to deter pests is a longstanding practice done by many gardeners. The most common misconception is that the strong fragrance of some herbs such as chives, rosemary, and garlic work to deter the pests that don't like those strong aromas. More accurately, companion planting in the garden can repel pests first by masking the smell of the host plant that the pest is looking for. Second, aromas attract predatory insects to naturally control the pest population. And third, companion planting works by confusing the pests. In her book *Plant Partners: Science-Based Companion Planting Strategies for the Vegetable Garden*, Jessica Walliser writes about the appropriate/inappropriate landings hypothesis, which states that an insect locates host plants "by making a series of landings on the foliage and 'tasting' it with the receptors on their feet." If multiple varieties of plants are all planted together, it disrupts the number of appropriate landings for the insect to determine which plants are a good host and could send the pest elsewhere.

Based on these three methods of repelling garden pests with plants, following are some herbs you can plant in your garden that can repel pests; however, this list is just a small sampling. Many more methods of fooling pests and using plants in the regenerative garden can be found in this chapter.

PLANTS

PLANT WITH BRASSICAS

Catnip   Nasturtium
Chamomile   Sage
Dill   Tansy
Hyssop   Thyme
Mint   Wormwood

PLANT WITH LETTUCE

Chives
Dill
Marigold
Mint

PLANT WITH SQUASH

Dill
Marigold
Mint

PLANT WITH POTATO

Catnip
Cilantro
Marigold
Tansy

PLANT WITH TOMATO

Basil
Borage
Calendula
Marigold
Nasturtium

PLANT WITH CARROT

Basil
Chives
Rosemary

PLANT WITH ASPARAGUS

Basil
Borage
Calendula

Polyculture for the Garden

Planting vegetables that are ready for harvest at different times, succession planting, and sowing multiple varieties of each vegetable are all part of growing a polyculture garden.

> Plant many varieties and species of plants in the garden for diversity and resilience.
>
> Interplant long- and short-season varieties, annuals and perennials, and groundcovers with tall plants.
>
> Plan for succession, save seeds, and mix plant families.

Monoculture cropping is done to manage large volumes of production of one crop at a time but it comes with many problems. Overproduction of one crop can lead to waste of what cannot be used or sold before it spoils. It invites pests and disease problems to thrive, and it depletes the soil of the nutrients the crop needs, and leads to the overuse of synthetic fertilizers, herbicides, and pesticides.

These problems are quickly beginning to outweigh the benefits and farmers are returning to polyculture, planting of many species and varieties of crops in order to more closely follow the natural processes of nature. These polyculture practices can work wonders in the home garden as well. Here are some tips for creating biodiversity of plants that will allow for more fruitful harvests and gardens.

- Sow seeds densely and eat what you thin.

- Plant several varieties. One or two may perform better than others.

- Reserve a minimum of 30 percent of the best plants for seed saving.

- Plant fast-growing edible plants along with plants that take a long time to mature.

- Plan varieties to ripen at different times to extend the harvest. Plant early- and late-season varieties.

- Sow seed varieties that mature at the same time every week or two so they are ready for harvest every week or two.

- Plant groundcovers below tall sun-loving plants.

- Research plant partners. Plant varieties that help each other grow, ripen, or stay safe from pests and disease.

- Plant deep-rooted plants with shallow-rooted plants.

- Mix plant families and species to discourage disease and pests from coming back in future years.

- Plant annuals around young perennials as they fill in the space and mature over the years.

PLANTS

How to Reclaim Weedy Soil

- Learn to identify weeds in the garden.

- Resist opening up the seed bank by disturbing the soil structure.

- Get to know wild plants for their benefits to the earth, people, and wildlife.

A weed is simply a plant that is growing where it is not wanted. If you have a bunch of rogue tomato or squash plants in your garden, they're called "volunteers." If you see them in a meadow or a wild space, they're "wildflowers." However, if you have hardy, native, or invasive plants growing where you didn't cultivate them, they are called "weeds."

Comfrey (*Symphytum officinale*) is a fast-growing, deeply-rooted wild plant that has many benefits to soil, wildlife, and people. It is often used as a compost booster or garden fertilizer. Both of those recipes can be found in my book, *Garden Alchemy: 80 Recipes and concoctions for organic fertilizers, plant elixirs, potting mixes, pest deterrents, and more*.

Weeds get a bad rap, and I'll admit, sometimes I'm not thrilled about them either (like when my bindweed tries to claim control over my garden). However, there are a few benefits to weeds:

- They are an important food source for insects, birds, and wildlife.

- Many are edible, tasty, nutritious, and medicinal.

- They cover bare soil quickly, holding in water and nutrition.

- They can also draw water and nutrients from deep in the soil via those long taproots that make them so hard to pull up.

- If you compost the weeds, the nutrients will feed your garden.

HOW TO IDENTIFY WEEDS

In most cases, the best way to identify if the little green leaves popping up in the garden are weeds is to let the plant in question grow and see what happens. As the leaves develop, plants become easier to identify. Weeds generally grow quickly and pop up where there are other plants already established, so they should quickly seem out of place.

If you identify a weed that looks beautiful when it's blooming that you want to keep for a short spell (such as those darling forget-me-nots), enjoy the flowers and if you don't want them to self-seed, pull the plant before it goes to seed.

HOW TO REMOVE WEEDS

The first step in removing weeds is to prevent unwanted wild plants from growing in the first place. Just think of it this way, every time you dig a dandelion root out of the lawn, you are creating the perfect opportunity for many other types of dormant seeds to germinate. Activities such as tilling soil and aerating the lawn open up the dormant seed bank that exists below the soil and signals to the seeds, which could have been happily dormant for ten years, that there is finally a space for them to grow.

Pick wild plants often and when they are small. Keeping up on the task daily or weekly will help eventually clear the weeds from an area. Weeding after rainfall or watering makes removing plants easier. Never let weeds go to seed, and don't add weed seeds to your compost bin. Most home compost piles do not get warm enough to sterilize weed seeds.

If you do have some unwanted wild plants growing, grab a tool such as a spade, soil knife, or hand weeder to dig into the soil and remove the roots of the plant. Fill the hole with compost and add some mulch to cover it to prevent the seed bank from germinating. Some weeds have deep taproots so the better you are at digging it all out, the less of a chance it will grow back.

In the vegetable gardens, it can be much easier to identify weeds. Generally, if a tidy row of similar-looking greenery is growing in an orderly fashion where you planted some seeds, it's likely cultivated. Random sprouts throughout the beds that don't match the new seedlings, however, are likely weeds.

Intensive Planting

Intensive planting can look like interplanting plants between vegetable garden rows, filling the spaces between perennial plants in garden beds, or planting the ground below trees and shrubs.

> - Add plants between vegetables and perennials to suppress weeds and increase biodiversity.
>
> - Interplant to extend the harvest and attract pollinators.
>
> - Interplant in all areas of the garden using the seven layers of a food forest, even if not all plants are food-producing.

For vegetable gardens, intensive planting uses the spaces between rows of large or long-to-mature plants to grow quick-to-harvest vegetable plants. For instance, planting radishes, arugula, spinach, and lettuces between rows of large or crops that mature in summer such as cabbage, cauliflower, and squash allows multiple harvests in the same plot of soil. You can keep interplanting throughout the seasons by adding fast-growing fall vegetables or cover crops between maturing summer vegetables three to four weeks before their harvest date. When the larger vegetables are harvested, the young plants between will have space to grow.

Intensive planting can also be a way to introduce more flowers and pollinator plants into the garden. Adding fast-growing annual flowers between perennials gives the garden more bloom time. Adding flowers such as *Alyssum* and *Phacelia* fill in spaces, suppressing weeds but also attracting beneficial insects such as predators and pollinators.

Intensive planting with cover crops and groundcovers such as clover allows even more weed suppression, attracts beneficial insects, and helps to fix nitrogen in the soil. Cover crops, mushrooms, vines, and epiphytes growing within rows of fruit trees or between the plants in a food forest is a good example.

PLANTS

Many seed companies have created eco-lawn alternatives full of hardy plant mixes that are perfect for lawn replacement as they require less mowing and water and can survive foot traffic.

- Sow eco-friendly lawn alternative seeds in your lawn.

- Allow lawn plants to go to seed before mowing.

- Grow a mix of plants that don't need supplemental water beyond annual rainfall.

Lawns don't have to be banished forever from the regenerative garden, but there are a few changes that need to be made to the orthodoxies we have been trained to believe about them. The question many homeowners first have about caring for their lawn is how they can get a lush, green turf to add curb appeal to their home. In a regenerative garden, a green lawn is absolutely possible and encouraged, as long as we adjust what green means. An eco-friendly lawn is a green biodiverse space that can host a large number of attractive, treadable green plants. In addition, these plants flower and provide food for wildlife, self-seed or spread to reduce the amounts of new seed we need to add to the lawn annually, and require less mowing and water.

You do not need to remove the turf grass to add eco-friendly lawn alternatives. Start by adding a mix of wildflowers and plants when seeding your lawn and see how it transforms from thirsty turf grasses to thriving, low-growing plants.

- Prepare the lawn by raking to remove any debris. Remove any unwanted weeds without digging too much. Opening up the soil allows dormant weed seeds to awaken and germinate, creating even more unwanted plants. Keep the weeding to a minimum and it will reduce the overall number of weeds you have to dig up in the future.

- For new lawns, seed three times in spring, for instance: March, April, and May. To fill in the lawn, seed twice in spring.

- Keep turf grass and clover lawns mowed no less than 2 to 3 inches (5 to 7.5 cm) tall to promote vigorous growth. Many wildflower lawns do not require mowing at all but will grow taller than is often preferred for a treadable lawn. Fertilize the lawn naturally by leaving the clippings on it after mowing. Let the plants grow a bit longer and go to seed before you cut them so they reseed.

Pretty five-spot (*Nemophila maculata*) is blooming in an eco-lawn and being visited by a sweat bee.

LAWN ALTERNATIVES

Common Name	Botanical Name
Baby Blue Eyes	*Nemophila menziesii*
Blue Star Creeper	*Isotoma fluviatilis*
Bugleweed	*Ajuga reptans*
California Poppy	*Eschscholzia californica*
Creeping Daisy	*Leucanthemum paludosum*
Creeping Thyme	*Thymus serpyllum*
Dwarf California Poppy	*Eschscholzia caespitosa*
Five-Spot	*Nemophila maculatea*
Hard Fescue	*Festuca maculata*

Common Name	Botanical Name
Johnny Jump-Up	*Viola tricolor*
Kidney Weed	*Dichondra micrantha*
Micro Clover	*Trifolium repens var. pipolina*
Periwinkle	*Vinca minor*
Sheep Fescue	*Festuca ovina*
Strawberry Clover	*Trifolium fragiferum*
Sweet Alyssum	*Lobularia maritima*
White Clover	*Trifolium repens*
Yellow Daisy	*Chrysanthemum multicaule*

It's completely natural and healthy for grass to turn brown in the heat of summer. Grasses do not like heat and will go dormant in very hot weather to protect the plants. Some lawn alternatives will also go dormant in summer but many will not. Planting for diversity will allow your lawn to better survive drought periods. Save water and allow the plants to thrive by supporting their natural cycle, not preventing it.

One simple test you can try is to go without watering your lawn for a whole year. Once you do this, you will quickly learn if the lawn grass is the right choice for your area. If the lawn can't survive without a lot of added water throughout the seasons, then think about changing it.

If you must water to establish a new lawn, limit it to 1 inch (2.5 cm) of water once a week to promote deep roots that will require less water and maintenance in the long run.

PLANTS

Planting a Guild

A cherry tree guild with wildflowers and herbs.

Guilds are collections of plants where each provides functions to work together as a system. As a group, they support and enhance the growth and productivity of the other plants. Think of a guild as a mini plant community. Plants in the guild can perform one or many functions. Guilds are often planted to surround fruit trees, but they can be built around multiple different kinds of plants. For instance, the Three Sisters method of planting beans, corn, and squash together is a guild common to the tribes of the Haudenosaunee Confederacy as well as many other Indigenous tribes of Canada and the United States. These three plants have a symbiotic relationship: the beans climb up the corn stalks while fixing the nitrogen in the soil for the corn, and the squash keeps the soil cool and reduces water evaporation while its blossoms attract pollinators. In Mexico, a similar agricultural method is called a Milpa where corn, beans, and squash fields are planted with other crops such as chilies, sweet potato, jicama, melons, amaranth, mucuna, avocado, and huitlacoche (known as corn smut, a fungus that grows in the corn and is considered a delicacy). Often a dozen or more plants are grown together, each supporting the others and being supported.

- Partner plants together to benefit one another.

- Build larger guilds with many plants that sustain and support each other.

- Move beyond food guilds and plant for wildlife, medicine, and soil remediation.

A fruit tree guild is designed with the fruit tree at the center of a round garden surrounded by fruit bushes and shrubs, pollinator plants, insectary plants, nitrogen fixers, groundcovers, and mulch plants.

The key components of a guild are:

1. **Food:** These are the edible crops that we use for ourselves and for wildlife. These can be fruits, vegetables, herbs, nuts, seeds—whatever we would grow and harvest as a food source or as medicine.

2. **Fixers:** The fixers are the nitrogen-fixing plants and organic matter that we add into the soil in order to replenish what is being used by the other plants. These plants also make nutrients in the soil available to the other plants in the guild. Often these are the nitrogen-fixing plants listed in the green manure project, but they can also be plants that add to the organic matter and nutrients in the soil when their leaves fall and become compost. As that organic matter breaks down, the nutrients are released into the soil and available for the other plants.

3. **Miners:** Miners cover those plants with deep taproots such as comfrey and dandelion that reach down and mine minerals to bring it up into their leaves, making them available at the surface level. When the leaves fall from these plants or they are composted by chop-and-drop, the nutrients become are available to plants with more shallow roots. Miners can also refer to what lives below the soil such as tree roots, root crops, worms, ants, termites, beetles, and digging mammals.

4. **Groundcovers:** Groundcovers act as a natural mulch to prevent moisture evaporation and suppress weeds. Cover crops can also be used to help prepare the soil for a new guild when it is first planted as discussed in the green manure project.

5. **Climbers:** Climbers make the most use of vertical space by climbing the supporters, which helps increase diversity and yield in a smaller footprint.

6. **Supporters:** Supporters are plants with thicker stems or trunks that support climbers. Tree trunks, corn stalks, and sunflowers are all great support plants. Trellises and obelisks can also be added as non-living supporters.

7. **Protectors:** Plants that attract predatory insects or repel pests act as the protectors of the garden. Thorny or sharp plants can also be added to deter pests from the garden.

8. **Pollinators:** Finally, plants that attract pollinators to the garden or act as pollinators are essential to include in a guild. Bird-, bee-, and butterfly-attracting plants bring pollinators to the garden and will help increase the yields of food plants. Additionally, some plants may need multiples in either the same or a different variety in order to be pollinated, and they should also be included in the guild.

Guild Recipes

BEES AND FRIENDS GUILD

Linden	*Tilia* spp.
Artichoke	*Cynara cardunculus* var. *scolymus*
Lovage	*Levisticum officinale*
Bee Balm	*Monarda*
Oregano	*Origanum vulgare*
Chives	*Allium schoenoprasum*
Borage	*Borago officinalis*
Clover	*Trifolium* spp.
Strawberry	*Fragaria*
Spring Bulbs	various

APPLE TREE GUILD

Apple	*Malus domestica* or *Malus pumila*
Raspberry	*Rubus idaeus*
Gooseberry	*Ribes grossularia*
Rhubarb	*Rheum rhabarbarum*
Nasturtium	*Tropaeolum* spp.
Sunflower	*Helianthus annuus*
Strawberry	*Fragaria x ananassa*
Borage	*Borago officinalis*
Mint	*Mentha* spp.

FRUITS, SHOOTS, AND ROOTS GUILD

Pear	*Pyrus* spp.
Blackcurrant	*Ribes nigrum*
Nanking Cherry	*Prunus tomentosa*
Asparagus	*Asparagus officinalis*
Beetroot	*Beta vulgaris*
Onion	*Allium cepa*
Marigold	*Tagetes* spp.
Radish	*Raphanus sativus*
Parsley	*Petroselinum crispum*

HERBAL MEDICINE GUILD

Witch Hazel	*Hamamelis* spp.
Elderberry	*Sambucus* spp.
Rose	*Rosa rugosa*
Lemon Balm	*Melissa officinalis*
Chamomile	*Matricaria chamomilla*
Peppermint	*Mentha × piperita*
Yarrow	*Achillea millefolium*
Thyme	*Thymus vulgaris*
Echinacea	*Echinacea* spp.

SOIL FIXERS GUILD

Mimosa/Silk Tree	*Albizia julibrissin*
Goumi	*Elaeagnus multiflora*
Comfrey	*Symphytum officinale*
Lupin	*Lupinus* spp.
Sunchoke	*Helianthus tuberosus*
Nettles	*Urtica dioica*
Clover	*Trifolium* spp.
Dandelion	*Taraxacum officinale*

How to Save Seeds

- Collect seeds from the garden for future planting.

- Plant heirloom seeds with seed saving in mind.

- Share saved seeds with neighbors and friends.

Seed saving helps you create your own food and plant security by cultivating the source to grow future gardens. Seeds can be saved from any plant that produces seeds; however, some plants are best propagated by cuttings or division. The seeds from trees, for instance, take many years to produce a mature plant. Plus, only heirloom seeds will reliably produce plants and fruit that are true to the parent plant. Heirloom seeds have been saved and passed on for many generations and each characteristic of the parent plant passes down to the next generation. By contrast, hybrid seeds come from cross-pollinating the strong characteristics of different varieties and gaining the benefits of that hybridization. This is also an ancient practice and one that plants do naturally when cross-pollinating in the wild, but the saved seeds may not grow true to the parent plants.

Saving seeds is a pretty simple process and can be done using the following methods.

- If the seeds inside fruit are not covered by any flesh and easily come out cleanly (such as bold peppers), simply cut the fruit open, remove the seeds, and allow them to dry on a paper towel.

- If the seeds inside the fruit are covered in flesh (such as **eggplant or squash**), wash the seeds by soaking them in water overnight. The next day, remove the seeds that have settled cleanly on the bottom of the container in which they've been soaking. Place these seeds on a paper towel to dry.

Dried poppy seedheads

Sunflower seeds ripening in the garden.

Ornamental allium seedheads

Plant different varieties of the same species apart to avoid accidental hybridization. Stagger plantings of different varieties of the same species so that they flower at different times. This will help spread the timing of seed collection.

- **Tomato seeds** can often successfully be saved using the last method; however, you'll have even more success if you allow the seeds to ferment. Scoop the seeds into a Mason jar along with the gel-like membrane surrounding them. Add enough water to cover the seeds and cover the jar with a paper towel secured by the ring lid. Label and allow the seeds to ferment until there's mold growing on the top of the mixture, the seeds have released from their membranes, and dropped to the bottom of the water. Rinse the seeds and dry them on a paper towel.

- Some seeds are best allowed to dry in place in the garden. **Peas and beans** can be allowed to dry in their pods before being picked and stored. **Poppy and other flower seeds** allowed to dry in the garden will rattle when the seeds are ready. To harvest, place a paper bag over the seedhead before picking it so that the seeds fall into the bag.

- **Strawberry** (*Fragaria × ananassa*) seeds grow on the outside of the fruit. In general, strawberries do start well from seed, but they also set up a lot of runners, which can easily be planted. **Alpine strawberries** (*Fragaria vesca*) are a smaller wild-tasting variety with a mounding habit. They are quite easy to start from seed. Collect the seeds by rubbing a ripe alpine strawberry on a paper towel and allowing it to dry. Carefully remove the seeds from the paper towel and store in a seed envelope.

Nigella has pretty seedheads that look good in dried flower arrangements.

- Design your food garden as a forest.

- Include as many of the components of a food forest as you can fit in your space.

- Determine the balance of plants that makes the garden regenerative without human input beyond harvesting.

Until recently, ancient pockets of food forests that exist within tropical rainforests have been seen by biologists as accidental, but there is increasing evidence that these forest gardens are a result of deliberate cultivation by Indigenous Peoples. These forest gardens are filled with a variety of food, layered, interconnected, and self-sustaining. In my home province of British Columbia, Canada, there's evidence of food forests that were planted more than 150 years ago.

A food forest is not just an efficient use of space to grow food but it is also a fully self-sustaining ecosystem. The most impressive feature is that once planted, a food forest can be walked away from for many years only for the gardeners to return and find it flourishing with a large amount of food. Food forests can be grown in many climates including dry lands, the tropics, temperate climates, and more forested regions.

This food forest has tall overstory trees to the north, a second layer of fruiting shrubs such as currants, and a sunny center plot for vegetable and herb plants.

The similarities of a food forest to a guild are many. The plants in a food forest are made up of a number of guilds all working together to support and feed one another in many different ways. The words "guilds" and "food forests" are often used interchangeably, but a food forest is designed specifically to produce the highest amount of consumable byproducts such as food, medicine, and craft materials in a small footprint. Of course, food forests are also designed to build soil and attract pollinators, creating a cohesive eco-system that is regenerative and long-lasting without the need for human effort to main-tain it.

Food forests are an ecological community all working together to support one another to grow and thrive. This type of ecosystem thrives with biodiversity, including plants, animals, and insects. The plants in the food forest are characterized by the layers in which they are growing. Listing them from sky to soil, here are the components that make up the structure of a food forest:

OVERSTORY TREES

The overstory is made up of the large foun-dational trees around which the food forest is based. The size of an overstory tree is what sets the overall footprint of the food forest. Majestic, 100-foot-tall (30-meter-tall) nut trees can create a wide footprint for plant-ing, while smaller fruit trees that reach 25 to 30 feet (7.6 to 9.1 meters) can be planted in groups. Keep in mind that the tallest canopy trees can be difficult to harvest properly. Much of the food grown will be left for wild-life and compost, but when it's mature, the tree will still provide plenty to eat as well.

UNDERSTORY TREES

Understory trees are shorter-growing but also can tolerate a little bit of shade because they are planted beneath the overstory trees. Typically these are native trees, such as dogwood and redbud, or ones that tolerate shade such as filbert and pawpaw. These trees can also be planted with succession in mind as they are ones that leaf and fruit earlier to take advantage of the sun before the overstory trees create shade.

SHRUBS AND BUSHES

The shrub layer is filled with perennials that have multiple woody stems that generally grow about 2 to 12 feet (0.6 to 3.6 meters) high. These are often berry plants that also provide wildlife habitat and wind protection.

EPIPHYTES

Epiphytes are not usually included as a layer in traditional food for people; however, they do provide food sources for wildlife and medicinal sources for humans. An epiphyte is a plant that doesn't grow in soil but instead attaches itself to the surface of another plant and gets its nutrients and water from the air. Moss and lichen are examples of epiphytes that grow well in shade and provide an important food source for wildlife. They also contribute to the bio-mass of the soil as they fall and decompose. Fallen lichen such as Usnea can be harvested as a medicinal herb.

Inoculated mushroom logs on the floor of a food forest.

VINING PLANTS

Climbing plants such as vines and bines are a way to take advantage of the vertical space of a garden to produce more food products. Some annuals (such as beans) produce lots of food and naturally die back after food production. Other plants (such as hops) are also productive but they can easily take over if not properly managed.

HERBACEOUS PLANTS

The herbaceous layer below the woody layer is filled with annual, biennial, and perennial plants. They're either self-seeded or grow back from their roots but, either way, they die back each year. Herbaceous plants vary in their sun and shade needs so plant the sun-lovers on the sunny side of the food forest and plant the shade-lovers on the shadier sides. This layer is essential for food production as well as creating organic matter that mulches and feeds the soil.

MUSHROOMS

Another layer that is not typically included in traditional seven-layer food forests is mushrooms. However, they're both nutritional and medicinal fungi that grow well in forest environments. Introducing mushrooms into your food forest is a great way to include another layer of consumables that will greatly benefit the entire ecosystem.

GROUNDCOVERS

The groundcover layer consists of herbaceous annuals, perennials, and biennials that grow low to the ground (such as strawberries, chickweed, clover, and herbs) but can also include ground vines (such as squash and sweet potatoes).

ROOT PLANTS

Root plants allow us to improve soil structure by aerating the soil but also to grow nutritious food crops just below the soil's surface.

PLANTS

DIY Trellis

The beauty of this trellis is that it can be made from just one material, is fast to make, and once you have the technique down, you can craft a trellis right in the garden whenever one is needed. It can be made with purchased or foraged materials, and it lasts many years.

- Make a single-material trellis to support climbing plants.

- Use found materials.

- Reuse the trellis many times until it is no longer sturdy, and then recycle the wood into a hugelkultur or garden bed.

There are many options for materials such as twine, reclaimed wood, untreated pallets, wire, and more. My favorite design is this pressure-held trellis that I used to make children's gardens. Straight branches or bamboo stakes are all that is needed to make a surprisingly sturdy trellis to support

THE REGENERATIVE GARDEN

82

lightweight climbers (such as peas, beans, cucamelons, and clematis). Using thicker branches will create more support, but it will also create more space between the branches, which would be fine for large climbers (such as evergreen clematis, cucumbers, and hops).

This smaller pea trellis is spaced 4 inches (10 cm) apart with fairly thin bamboo that measures 5 feet (1.5 meters) high.

MATERIALS

The first thing to consider is the plant you would like to use trellis to hold up. Ensure that the wood you are choosing is sturdy enough to last; found branches should be straight and strong. If the wood is dry or rotting and breaks easily in your hands, add it to your hugelkultur and find different branches for this project. Bamboo is a good choice for this project as it is a very strong and hard material. If you have some available, galvanized metal and reclaimed wood from fences or pallets can work as well.

MAKE IT!

1 Place poles about 6 to 12 inches (15 to 30 cm) apart, depending on the size of what you are working with, by firmly staking the ends into the ground.

2 Cut the bamboo 6 to 12 inches (15 to 30 cm) longer than the distance between the two end poles. Weave the length of bamboo through the stakes by passing it alternatively behind and in front of each stake. Slide the pole down through the stakes to about 6 inches (15 cm) from the bottom of the stakes.

3 Repeat this step with another pole but alternate how the pole weaves front to back with the first pole. Continue this process until you have a sturdy trellis that has the spacing you are hoping for.

You will be surprised at the strength of this trellis and how easy it is to make. If you have access to a lot of sturdy and straight branches, you could also make this structure from those. It will look beautifully rustic and reuses a waste material for one last hurrah before it returns to the soil.

PLANTS

83

Keep heavier fruit off the ground with a lean-to trellis made from a wood frame and cross-hatched wire.

The Art of Espalier

Growing espalier trees in the home garden is a decorative way to grow more abundant tree fruit in small spaces. Training trees, shrubs, and even groundcovers on vertical structures can make for decorative living walls that create interest.

- Use espalier to decorate and add plant life to fences and buildings.
- Plant multiple varieties of each fruit for diversity.
- Grow espaliers in a guild or food forest.

PLANTS

85

Training and pruning fruit trees to grow along walls or fences keeps the fruit at an easily accessible height and turns an otherwise standard tree into a garden showpiece.

Espaliers can be fruit trees or ornamental, evergreen, or deciduous trees. This *Euonymus fortunei* (wintercreeper) was trained as an espalier up a privacy screen for four-season interest.

ESPALIER SHAPES

There are many different shapes of an espalier: cordon (branches straight out to the sides), fan (branches fanning up and to the side), candelabra (like a cordon but the branches turn at a right angle to form the shape of a candelabra), lattice (multiple trees with crossing branches), and "Y" shapes.

The simplest shape to start is the cordon to span a fence or cover a larger area with few trees. For denser plantings, such as in a heritage espalier orchard, the lattice shape allows many trees to cross over one another in rows.

A heritage apple orchard can be planted in a lattice pattern. This makes a very easy-to-manage orchard with the fruit growing prolifically on shorter trees that are simpler to prune and harvest.

Espalier fruit trees can often be purchased with four to six different fruit varieties grafted onto a dwarf rootstock. Grafting is the process of attaching a branch to the tree so that they grow together as one. Since the trunk of a tree supports the rootstock, it determines the overall height of the tree. Then the branches of different trees are grafted to produce varied fruiting branches. Often these are compatible varieties of the same type of fruit (apple, pear, Asian pear, and cherry, among others) that support the pollination of each other.

The espalier in my garden has five different apple varieties grafted onto a dwarf apple rootstock. I have space for only one tree, so I chose a tree with different varieties of apples that flower and fruit at varying times throughout the season. The varieties are selected to support one another so that cross-pollination can occur.

HOW TO PLANT AN ESPALIER

The optimal time to plant fruit trees are in winter or early spring when they are dormant. Here's how:

1 Dig the tree into the soil as soon as the soil is workable for the year.

2 Create a large hole that is twice as wide, but just as deep as your rootball.

3 Add well-rotted compost to the hole.

4 Position the tree so that the base of the trunk at the root flare (just where it begins to widen) is at the soil line. Plant any deeper and the roots will grow upwards; plant too high and roots will be exposed. Fill in the hole with soil and water well for its first year until it's established.

5 Don't forget to pick the right place for your tree. Most fruit trees love sun, so a nice sunny spot will give you the best fruit.

6 Plant a row or multiple trees depending on the spacing suggested for each variety, usually between 12 to 20 feet (3.7 to 6.1 m) apart.

HOW TO PRUNE AND TRAIN AN ESPALIER

First, determine the pattern you want and look for a young tree that has that basic shape. Then:

- Remove any branches that don't fit the pattern or that suffer from one of the four Ds of pruning: dead, diseased, damaged, or dying.

- Build a structure of wires between posts to support the shape or attach the branches to an existing structure such as a fence. Vertical wires or wood slats should be approximately 18 inches (45 cm) apart.

- Use a soft covered wire or ribbon that can be retied when the branches grow. Be sure not to choke the branches with too-tight ties.

Before pruning

After pruning

ESPALIER MAINTENANCE

Annual pruning will keep an espalier neat and productive. For deciduous trees, prune after it has set flowers or fruit for the year but before it goes dormant. In fruit trees this is often a midsummer pruning. Remove any branches that are starting to get long and leave plenty of buds where the cuts are. This will ensure that leaves, flowers, and fruit grow close to the branches.

If the tree produces a lot of new growth after this pruning, prune again before dormancy.

Dormant pruning in winter helps maintain the shape of the espalier and can allow you to hone the shape of each branch. Keep in mind that overly vigorous winter pruning can create a surge of fast-growing watersprouts in spring that will need to be removed. It's best to prune with reserve during a tree's dormancy.

In the growing season, a light, selective pruning can help thin some of the flowers and forming fruit. By pruning any long branches, watersprouts, or heavy clusters, the tree's energy will go to the remaining flowers, fruit, or leaves closest to the trunk and branches.

Grow a Bee Border

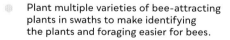 Plant multiple varieties of bee-attracting plants in swaths to make identifying the plants and foraging easier for bees.

Plan for succession to ensure there is a long blooming period.

Plant native bee plants that are drought tolerant and adapt easily.

I've been known to spend time with an artist's paintbrush hand-pollinating plants to improve fruiting. All the while thanking the bees around me for the hard work that they do to make my garden bountiful. Creating a border of bee-attracting plants around your vegetable garden helps bring in a workforce of pollinators and gives you a better harvest.

Once you attract bees and other pollinators, they search for more flowers. A few small steps can help you create a wonderful bee border that will be welcoming to bees, allowing you to sit back and enjoy the hustle and bustle as they work.

Here are some helpful tips for creating a border:

- A simple and inexpensive idea for a bee border is to scatter bee-blend wildflower seeds and loose soil around the garden. A mix of annual and perennial seeds will give the longest bloom times. Look for seeds from local seed companies that specialize in bee and pollinator blends for your area. That way you'll start to attract the largest number of wild bees naturally.

- Plant multiples of each variety so that bees see large swaths of flowers to forage in.

- Plan your bee plants so that they are blooming from early in the season until very late.

- Plant flowering native plants to attract and maintain wild bees.

- The majority of native bees are ground-dwelling bees so it's important to ensure care when maintaining the garden. Leaving mulch in place and only digging when absolutely necessary will allow ground bees to stay undisturbed in their homes.

- Add a bee bath. Fill a shallow dish of water halfway and place a few pebbles to sit above the water line so bees can land safely and drink while they forage and nest.

DON'T FEAR THE BEES!

Bees are gentle creatures that only sting to defend themselves from harm such as being grabbed or trampled. Wasps can be a bit more aggressive, but mostly just in the pursuit of your BBQ dinner. Unless you provoke them or threaten their nest, wasps won't start attacking you. Remember that bees and wasps don't get angry; they simply defend themselves. Show them respect and care and you should not get stung.

Pollinators are hardworking garden helpers that you only need to pay in pollen and nectar.

Without pollinators, where would we be? If you've ever had a less-than-impressive harvest it could be because you didn't have enough bees.

PLANT THESE FOR THE BEES

One way you can figure out what plants to choose for your garden is to look for the bees and let them tell you what plants they like. Observe which plants they are buzzing around at a community garden, meadow, or even the garden center. Instead of planting many different varieties, plan to have a number of the same variety of flowers because bees are attracted to larger expanses of one kind of flower.

ANNUALS

Aster	*Asteraceae*
Cockscomb	*Celosia*
Clover	*Trifolium*
Spider Flower	*Cleome*
Beeblossom	*Guara*
Marigold	*Tagetes*
Millet	*Pennisetum glaucum*
Nasturtium	*Tropaeolum minor*
Poppy	*Papaver*
Perennial Salvia	*Salvia*
Statice	*Limonium sinuatum*
Sunflower	*Helianthus annuus*
Sweet Alyssum	*Lobularia maritima*
Zinnia	*Zinnia elegans*

HERBS

Anise Hyssop	*Agastache*
Borage	*Borago officinalis*
Catmint	*Nepeta*
Pot Marigold	*Calendula officinalis*
Dandelion	*Taraxacum*
Fennel	*Foeniculum vulgare*
Lavender	*Lavandula*
Lemon Balm	*Melissa officinalis*
Mint	*Mentha*
Oregano	*Origanum vulgare*
Rosemary	*Salvia rosmarinus*
Sage	*Salvia officinalis*
Thyme	*Thymus vulgaris*

PERENNIALS

Ornamental Onion	*Allium*
Anemone	*Anemone hupehensis*
Bee Balm	*Monarda*
Black-Eyed Susan	*Rudbeckia hirta*
Clematis	*Clematis*
Montbretia/Falling Stars/Copper Tips	*Crocosmia*
Crocus	*Crocus*
Dahlia	*Dahlia*
Coneflower	*Echinacea*
Geranium	*Pelargonium*
Globe Thistle	*Echinops*
Hollyhock	*Alcea*
Beardtongue	*Penstemon*
Autumn Joy, Stonecrop	*Sedum*

4
Climate
Creating Harmony

Learn about your unique microclimate and match plant choices for success.

Create structures to protect plants and people from climate extremes.

Capture, store, and use energy produced naturally.

Let's talk about the weather, shall we? As gardeners, we are constantly following the weather patterns to best plan and plant our gardens. In order to give us some standardized information, we look at climate, the long term-weather average in your area. However, with the changes to climate that are rapidly occurring around the world, following a thirty-year historical average of data is no longer a tool that can be counted on. Plus, we all have unique microclimates that exist in our gardens, often more than one.

Just as I've designed the projects in this book to span many different soil types, water needs, and plant preferences, this chapter will help you find the ways to work with the climate in your unique garden ecosystem. Despite vast differences from city to city, province to state, or country to country, there are a few key elements of climate that we can work with in our gardens: temperature, wind, sunlight, and water.

The temperature and sunlight levels in your home garden determine the plants you can grow and how long your growing season is. The projects in this chapter are intended to spark ideas on how to harness the energy of the climate and design systems that work within your unique microclimate. In colder regions, you can capture heat in thermal masses or greenhouses. In warmer regions, you can cool plants using shade and water. There are ideas in this chapter for collecting heat energy, windbreaks, planting for shade, and even creating a privacy screen for sound and sight dampening.

Umbrella Greenhouse Cloches

Cover seedings with an umbrella cloche in early spring.

- Create a mini greenhouse from a clear umbrella.

- Use multiple umbrella greenhouses around the garden to extend the season and speed ripening.

- Thoroughly dry and lubricate the umbrella mechanism at the end of seasonal use, and store in a dry location to prolong the life of its metal hardware.

Greenhouses are a wonderful way to capture and store energy such as heat and light to extend the growing season in cooler climates. They can also protect plants from birds, insects, and disease that is air- or water-borne. Make a simple miniature greenhouse

Lettuce and radish seedlings are growing early in the season under cover.

An early spring salad garden is ready to harvest.

by using a clear umbrella as a cloche over a half-barrel planter. These planters are wonderful for starting seeds early in the season and their large size allows you to grow a wide variety of plants while extending the growing season.

Prepare a wine barrel for planting by drilling drainage holes and then setting landscape fabric over the holes to prevent soil from filling them in.

For annual crops and small shrubs, fill the bottom of the barrel with chopped pieces of logs and branches. For trees or deep-rooted plants, skip the branches and logs. Fill the barrel with a potting soil mix of compost and organic materials such as coconut coir, rice hulls, and/or perlite.

Plant seedlings or sow seeds, water well, and cover with the umbrella.

Rain will not water the seedlings, but more moisture will be retained in the greenhouse. Manage water to ensure that seeds or seedlings do not dry out and vent the greenhouse often to prevent mold growth. On warm days, remove the cloche to regulate temperature. As the weather warms, begin hardening off the seedlings by removing the greenhouse cloche for an increasing amount of time each day until you no longer need it.

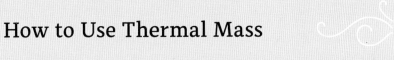

A concrete fountain is a lovely place for heat-loving and drought-tolerant succulents to naturalize.

- Identify structures that hold heat and plant heat-loving plants near them.

- Use recycled materials to build a thermal mass such as an herb spiral raised bed.

- Integrate irrigation or water catchment such as wicking beds into thermal mass gardens.

A thermal mass is any structure that captures, holds, and radiates heat to warm the soil, plants, and air surrounding the mass. The most common example is a concrete sidewalk. It warms up from the sun's energy and holds heat, warming the ground and anything that walks on the concrete.

CLIMATE

A rock wall can capture, hold, and radiate heat in summer.

In highly urbanized areas, the lack of plant material such as grass and trees surrounding buildings, roads, and sidewalks can create a "heat island" where temperatures are much hotter than they would be if the area were forested.

We can use this idea to add heat islands to our gardens for heat-loving plants, and more thoroughly utilize the existing structures that act as thermal masses such as stone walls, pathways, and buildings. Simply planting a citrus tree near a south-facing wall in a cool-climate garden that would normally not have the right conditions for citrus to grow can increase the temperature enough to produce a fruitful harvest.

IT'S AS SIMPLE AS BLACK OR WHITE

Just as the color of our clothing affects how much heat we absorb from the sun, we can choose materials or paint to control heat on our garden structures. Dark colors absorb and hold heat while light colors reflect it. Using dark stones for a retaining wall or dark paint on a building will warm the structure, while using light-colored materials or painting structures white or pastel colors will not hold as much heat. Measuring where sunlight falls in summer and winter and adjusting the paint colors of structures either to capture or reflect the light can help to reduce energy use inside buildings throughout the year.

Build an Herb Spiral

- Increase planting square footage by building an herb spiral.

- In cool climates, use materials that hold heat to increase the temperature of the soil. In warm climates, use materials that do not hold heat.

- Integrate irrigation or water catchments such as a pond at the base of the spiral.

This project was built by by Kristen Noble, Vegetable Product Manager with Harris Seeds in Rochester, New York. It took only 45 minutes to construct yet it's both beautiful and a long-lasting feature garden in the display garden and practical place to trial their herb seeds.

CLIMATE

103

MATERIALS

- Large level or long 2 x 4 with a smaller level attached
- String
- Stone retaining wall blocks
- Soil
- Garden hoe

MAKE IT!

1 Level the soil where you plan to construct the spiral bed. If the area is grassy or weedy, remove the vegetation and/or cover with a layer of damp cardboard. Chose a space with at least 5 to 6 feet (1.5 to 1.8 meters) width in all directions.

2 Use string to draw the shape of the spiral. Be sure to consider the size of the stones and allow enough space to plant between the walls of the spiral. Place the bottom of the spiral toward the north side of the garden area where it will receive less sun. Another option is to add a small pond at the bottom of the spiral to grow watercress, water chestnuts, or other pond plants.

An herb spiral is a type of space-saving structure that can stretch a bed that is 5 feet (1.5 meters) in diameter to 20 or 30 linear feet (6 to 9 meters). A typical herb spiral could be made from a variety of materials such as wood, gabions, or straw bales. In this case, using stone retaining wall blocks absorb and hold heat, which allows climate control and energy harvesting, making it an ideal structure for herbs with many different climate needs. For warm climates, choose wood or straw to prevent the soil from getting too warm.

The south-facing side of the spiral will get full sun, while the north side will be shaded and cooler. The east side gets more gentle morning sun, but water evaporates more quickly. The west side gets the harsher afternoon sun and heat but retains moisture in the morning. The top layers of the spiral will be drier, and the bottom will be wetter. The entire bed will have warmer soil than what's in the ground when the sun warms the stones.

3 When placing the first layer of stones, use a hoe to lower the soil level where the stones will sit so that they rest a few inches (centimeters) below the soil. This will help to keep the foundation stones in place. While doing this, be sure to keep your soil level. Place the first layer of stones around the path of the spiral.

4 Add layers of stone to increase the height of the walls of the spiral to be four stones high at the top of the spiral. For improved stability, place the top layers of stones over the joint of the two stones below.

5 Start the second layer of stones by overlapping the first stone at the outer edge. Start the third layer by overlapping the fifth and sixth stones of the second layer. Start the fourth layer overlapping the fourth and fifth stones of the third layer.

6 Add soil to the spiral during construction to help hold the stones in place.

7 Once you've built the spiral to your desired height, fill the bed and any gaps between the stones with soil. Water the bed to help the soil settle into the space.

8 Add drip irrigation around the spiral to allow easier watering.

WHAT AND WHERE TO PLANT AN HERB SPIRAL

NORTH BOTTOM
lettuce
salad greens

NORTH TOP
parsley
chives

EAST TOP
calendula
chamomile
sorrel

WEST TOP
sage
yarrow
fennel
salad burnet

WEST BOTTOM
strawberries

EAST BOTTOM
violets
pansies
cilantro
chervil

SOUTH TOP
echinacea, dill
rosemary, lavender
oregano, thyme

SOUTH BOTTOM
feverfew
basil

Reverse these cardinal directions if you live in the southern hemisphere.

How to Measure Light
(Sun and Shade Mapping)

Sun mapping is the process of watching the sun's path over a year, logging its path and how it changes throughout the seasons.

- Draw a basic map of your garden and then observe and record where and when the sun rises and sets for each quarter of a year.

- Draw the sun patterns for each month of the year in a sketchbook. Add rainfall measurements using the rain gauge from chapter 3.

- Plan plants and structures that best work with the sun pattern in your garden.

Are you wondering whether your garden is sunny, shady, part-sun, part-shade, or something else? The answer is that it is probably all of those things! Different parts of the garden as well as the different times of the year change how much sun directly reaches the plants in a day. In winter, the days are much shorter than in summer, so even in a fully sunny spot, the number of daylight hours is more indicative of a partly sunny spot. The sun will rise and set on a different path during each season. While it may be directly overhead at 12 p.m. in midsummer, it certainly won't be in that same place in spring and fall.

MAKE IT!

1 Draw a rough map of your garden; it doesn't have to be perfect or drawn to scale. If you mark the boundaries and main structures such as houses, outbuildings, trees, and ponds, you will have a sense of the basic layout. Next add the four cardinal directions (north, east, south, and west) at some place on the map. You can then make copies of this map and use it to draw the sun each season or redraw the map in a journal.

2 Spend a few days watching where the sun rises and where it is at various times throughout the day (early morning, midmorning, noon, midafternoon, late afternoon, evening); then draw a circle on the map to show where the sun spreads. You will have six overlapping circles at the end of the day. Color the circles with yellow and orange pencils: yellow to indicate gentler sun and orange to indicate more intense sun.

3 Use these maps when planning your garden for sun or shade conditions. Parts of your garden that get sun all day long will enjoy the hardiest, sun-loving plants or it will need trees planted to provide shade. Some areas might be more shaded by buildings and trees at different times of the year, but they get full sun in other times. This will help you match the plants in those areas to the sun and shade needs they have naturally.

How to Build a Windbreak

Windbreaks do not need to be tall or permanent to be effective.

- Provide protection and comfort by installing a windbreak.

- Use plant material such as a row of trees or a hedgerow as a windbreak.

- Include native plants, fire-resistant plants, and wildlife hedge plants as part of a windbreak.

Wind is a beautiful function of our gardens. It can cool the air, shape the land, pollinate plants, and help propel birds and insects. That being said, it can also be too much of a good thing. If you have open spaces where cleared trees have left the garden exposed, windbreaks can help provide comfort and protection.

A windbreak is any structure that provides protection or shelter from wind. It could be in the form of a hardscaped structure such as a fence or wall, temporary materials such as hurdles or burlap, or landscaped plant material.

A line of trees at the edge of a property can be a valuable comfort measure to protect from wind. Choose fire-resistant shrubs or trees in areas that are prone to wildfires.

Planted windbreaks can be fast-growing annuals such as giant sunflowers, perennials such as pampas grass, or rows of shrubs or trees. When logging the microclimate conditions of your garden, note any regular wind patterns that occur.

A windbreak will protect between 5 to 10 times its height beyond the windbreak. For instance, a 5-foot-tall (1.5-meter-tall) hedge protects the next 25 feet (7.6 meters) to 50 feet (15.2 meters) beyond the windbreak.

PLANT FIRE-RESISTANT TREES IN WILDFIRE-PRONE AREAS

Wind feeds and drives fire while oxygenating the air, so including fire-resistant trees such as oak, walnut, or hawthorne can help slow wildfires. More information on fire-resistant plants can be found at https://www.firefree.org/fireresistantplants/.

HURDLES AS WINDBREAKS

Isle of Man author and founder of LovelyGreens.com, Tanya Anderson, discovered the history of hurdles as they are used in the garden. While hurdles are typically thought of as a track and field sport, they weren't invented for jumping over! After seeing hurdles on a garden tour, she shared this story.

Hurdles are "moveable sections of fence that you can use for various farming and gardening purposes. One of them is protecting crops from wind and weather. Manx (Isle of Man) farmers used to use hurdles to protect winter cauliflower crops. They're not used commercially these days but I'm sure garden-grown brassicas would do better with a little extra protection."

This willow hurdle is protecting the squash, parsley, dahlias, and oca growing on the other side.

How to Create a Privacy Screen
as a Visual and Noise Buffer

- Use shrubs, trees, or a trellised climbing plant to create a green privacy screen.

- Use recycled building materials.

- When it comes to neighbors, build longer tables, not higher fences.

An espaliered evergreen climber on a wood trellis makes for a quick-growing, small-space solution that looks beautiful in all four seasons.

While I love to use trees and large shrubs to frame the perimeter of the garden, they can take up a lot of space. Privacy screens as a garden feature take up only a small footprint and can be effective as an environmental buffer. They can reduce noise and smells from adjacent properties as well as cover unsightly structures. That being said, I dislike the common term "privacy screen". If you do not have a need for an environmental barrier from neighboring properties, enjoy the sharing that comes from connecting with neighbors. The community chapter has a number of ways to do develop bonds within the community.

The most important part of building any structure is to make sure that its foundation is strong enough to prevent it from tipping or blowing over on a windy day. A good rule of thumb is to dig each posthole at least one-third the depth of the fence's height. For this project the privacy screen is 11 feet (3.4 meters) tall, so the postholes need to be at least 3 feet 6 inches (1 meter) deep.

Many fence posts are built by placing the actual post in the posthole and pouring concrete around it, but for this project I chose to seat the posts onto post saddles instead to help keep them out of the soil and water. This will reduce the risk of the posts rotting prematurely.

Dimensions:
11 feet high x 8 feet wide
(3.5 meters x 2.4 meters high)
adjust for your space.

MATERIALS

- (3) 4x4x12 pressure-treated wood
- (2) 2x4x8 pressure-treated wood
- (15) 1x4x8 cedar wood
- (15) 1x2x8 cedar wood
- (1) 1x4x12 cedar wood
- (about 16) 60-lb. (27-kg) bags of concrete
- 12x12 concrete forming tubes
- 1½- and 3-inch (3.8- and 7.6-cm) deck screws
- (2) galvanized post saddles
- (8) ½x2-inch (1x5-cm) galvanized lag bolts
- (4) ½x6-inch (1x15-cm) galvanized lag bolts
- (2) 6-foot (1.8-meter) galvanized angle iron
- Gravel

TOOLS

- Shovel or post hole digger
- Wheelbarrow or concrete mixer
- Circular saw
- Levels
- Hammer
- Drill
- Scrap 2x4 lumber
- Step ladder that will allow you to safely reach at least 12 feet (3.6 meters) high.

2

MAKE IT!

Prepare the Concrete Footers and Saddles

1 Dig two holes 7 feet 2 inches (2.2 meters) apart on center, 3 feet 5 inches (1 meter) deep, and 14 inches (34.5 cm) wide.

2 Cut a concrete forming tube into two 3 feet 8 inch (1.1 meter) lengths, place each tube into the post holes, and make sure that the top of the tubes are above grade.

3 Backfill around the outside of each tube with dirt.

4 Tamp down the bottom of each hole and pour in 3 inches (7.6 cm) gravel.

5 Pour about 6 bags of prepared concrete in each tube while periodically tamping the concrete.

6 Round the top of each concrete footer away from the center to prevent rainwater from pooling.

7 Insert post saddles into the center of each concrete footer. There should be about a 1-inch (2.5-cm) gap between the bottom of the saddle and the top of the concrete footer to prevent the post from contacting with water and soil and prematurely rotting. Ensure that the post saddles are level vertically, horizontally, and with one another.

8 Allow the concrete to cure.

Attach the Privacy Screen Posts

1 Cut two 4-inch x 4-inch x 12-foot (10-cm x 10-cm x 3.7-meter) pressure-treated posts to 11 feet (3.4 meters).

2 Seat each post onto the saddle and level them vertically.

3 Stabilize the posts with a scrap 2x4 anchored both to the posts and the ground to ensure that they stay level while completing the rest of the privacy screen structure.

4 Use 2-inch (5-cm) galvanized lag bolts to firmly attach the posts to the saddles.

Stabilize the Posts (optional)

The posts should not sway or lean when you push on them. If you feel that the posts are not stable enough, then follow these steps to provide additional stability.

1 Prepare two concrete footers about 2 feet (0.6 meter) deep by 12 inches (30 cm) wide, directly behind each fence post.

2 Insert two 6-inch (15-cm) galvanized lag bolts into the ends of each 6-foot (1.8-meter) galvanized angle iron.

3 Insert the angle iron (lag bolt first) deep into the concrete footer and attach the other end to the corner of each fence post using two 2-inch-long (5-cm-long) lag bolts.

4 Allow the concrete to cure.

Attach the Rails and Braces

1 Cut two 2-inch x 4-inch x 8-foot (5-cm x 10-cm x 2.4-meter) pressure-treated planks to fit the inside length of the two posts and attach them 6 inches (15 cm) from the top and 6 inches (15 cm) from the bottom of the posts.

2 Cut a 4-inch x 4-inch x 12-foot (10-cm x 10-cm x 3.7-meter) pressure-treated post into 4 equal pieces; trim the ends at 45-degree angles; and attach them to the inside corners of the privacy screen structure. This will help prevent diagonal swaying.

2

PLANT IT!

Finally, you can plant your espalier at the base of the structure and use garden ties to attach the branches to the lattice. Try adorning the screen with other decorative elements as well; shown here is a grapevine wreath to fill in the empty space at the top.

Install the Lattice

The lattice is made up of alternating strips of 1- x 2-inch (2.5- x 5-cm) and 1- x 4-inch (2.5- x 10-cm) cedar planks with a 1½-inch (3.5-cm) gap in between each strip. To make measuring the 1½-inch (3.5-cm) gap easy and consistent, use the wider part and extra 1- x 2-inch (2.5- x 5-cm) cedar plank as a spacing guide.

1 Cut a 2-inch x 4-inch x 12-foot (5-cm x 10-cm x 3.7-meter) cedar plank to 11 feet (3.5 meters) and attach it vertically to the center of the top and bottom rails. This center plank will help prevent the lattice strips from sagging.

2 Working from the top down, attach horizontally to the posts and vertical center plank the 1-x 2-inch (2.5- x 5-cm) and 1- x 4-inch (2.5- x 10-cm) cedar planks in alternating order with a 1½-inch (3.5-cm) gap between each plank.

To create a healthy ecosystem in our home gardens it's essential to add trees or protect those trees already growing in the space.

Reforestation for Cooling and Moisture Retention

One of the most disruptive things we've done to our landscapes is to remove all the trees. We've created barren, sunny plots of land in order to build cities, neighborhoods, and farmland. Trees are arguably the most important plants that we have on our planet. They feed the soil, create biomass, sink carbon, provide habitat for wildlife, provide food and medicine, and protect us from climate effects such as wind, sun, heat, and rain.

Trees infuse rainfall with nutrients from salts, minerals, and organic matter, which slowly drip to the ground. Annually, trees can give back a percentage of their biomass to build soil and feed surrounding plants. Trees provide shade that helps keep the ground cool, reducing evaporation and the need for supplemental watering.

This is why city design planners include many trees in urban and suburban spaces. A downtown core without trees becomes a concrete island. The sidewalks, buildings, and roads all attract and absorb heat, raise the overall temperature, and force evaporation of any groundwater. Planting trees is an effective way to capture water before it hits hot concrete and evaporates, allowing it to return to the soil and the plants while cooling the land and the surrounding air. The shade that is created by city trees reduces the infrared light from the sun that heats the roads and sidewalks.

If you're looking for a way to cool down your property and increase drought tolerance, then planting trees throughout your property is an excellent idea.

> Protect and foster trees.
>
> Plant trees.
>
> Plant more trees.

CREATE A SOCIAL NETWORK

Only recently has it become commonly accepted that trees are social beings that live in communities and communicate with one another. Through their roots and a symbiotic relationship with soil fungi, trees can exchange water and nutrients. Using an underground network of mycorrhizae (from the Greek words for "fungus" and "root"), mature overstory trees convert sunlight to produce food for the understory trees and saplings. Trees also use the fungal network to send alarm signals when there is a threat. Keep this in mind when you are designing your garden because a lone tree is a lonely tree. Plant trees together and allow them to live in a community as nature intended.

5
Ethics
Reducing Waste & Encouraging Diversity

 Reduce, recycle, and reuse materials, products, and energy.

 Maximize the use of each product before disposing of it. Consider how many different uses a material can have in its lifetime.

Consider the waste impact before acquiring any new item and seek out materials that can be repurposed.

If I'm being honest, I didn't fully think through the future of our beautiful earth when I had a child. As a young person, I felt more connected to my lifetime. Once a child graced my life, I started to think of all the lifetimes that are impacted by how we live. The state of our climate emergency shows how important it is to be conscience of our choices and how they affect the earth. We "live for today" and "focus on the present." Yet we have a responsibility as borrowers of this planet not just to the future generations of humans but to the future generations of the plant and animal species we share it with. We have a responsibility to the soil wildlife, the ocean creatures, and the declining insect populations not to continue taking without thought or care for how it affects others.

Our current reality is that a new set of ethics is what's now needed for our survival as a species. The earth will be just fine long after we are gone. Truly, the only option here is to respect our Mother Earth and thank her for allowing us to live here by getting along with our neighbors, the flora and fauna with which we share this beautiful place.

This may sound grandiose and dramatic—perhaps even over-whelming. Climate grief is real, and it can make folks feel hopeless. But if we act ethically and build regenerative systems, we are creating hope. This chapter focuses on reducing the waste that we create, reusing more of the materials that we acquire, and finding a symbiotic relationship with our land.

Our garden spaces can produce more food and flowers, more trees and shelter for wildlife, control climates for homes and outdoor spaces, and spread joy and beauty within our neighborhoods, all with less input from us. The time- and cost-savings alone should be attractive enough for us to want to make a regenerative garden.

The benefits of this work have a ripple effect. Each time we collect rainwater and use it to irrigate our vegetable garden, we are saving the effort of our communities and the land to purify that water and redeliver it to us. Every time we compost our garden waste to turn it into rich fertilizer, we save the energy of it being collected by our cities and composted and sold back to us. Every little step that we can take will make a bigger difference than is immediately evident.

Keyhole Bed

Top-down view

Cut-away view

- Build a keyhole-shaped bed to maximize planting space and limit pathways.
- Use recycled materials to build the structure.
- Irrigate with rainwater and include composting in the bed design.

The keyhole bed is a brilliant design that maximizes the amount of garden space and minimizes the number of pathways that it takes to access that space. It's a very efficient design that allows you to walk into the keyhole to tend to the garden plants that are in arranged 300-degrees around you in the center when you are standing in the "keyhole."

There are two types of beds that are commonly called keyhole beds. African keyhole beds, as shown in the illustration, are raised round beds with a path that leads to a center composting bin. These beds were designed in the 1990s to be accessible, self-sustaining kitchen gardens. The walls are tall enough to lean on plus the design provides easy access to food and an immediate place to compost the garden waste which feeds the surrounding plants. The layers of the beds can be built similarly to a hugelkultur, with logs at the base and layers of carbon and nitrogen, using whatever materials are available to fill the bed and create soil. The top 6 to 12 inches (15.2 to 30.5 cm) of compost can be planted directly.

In permaculture, a keyhole bed refers to any bed that has an efficient horseshoe shape with a path leading into the center. They sometimes have a composting bin in the center, but this limits the reach of some plants. In a small bed, you can spot compost and/or create a worm tower in the keyhole bed in order to compost in place without taking up that valuable center space.

In the book *Gaia's Garden* Toby Hemenway writes, ". . . for 50 feet (15.2 m) of planting, traditional rows of vegetables require 40 sq. ft. (3.7 sq. m.) of path, raised beds require 10 sq. ft. (0.9 sq m.) of path, keyhole beds require 6 sq. ft. (0.6 sq. m.) of path."

The frame of a keyhole bed can be made from wood, bricks, rocks, or really any other material that is available to make beds. If you're starting a new keyhole bed, then sheet mulching and hugelkultur concepts can be employed to build the soil base. The keyhole concept has to do with the shape and efficiency of the design.

A keyhole bed is a round, curved, or horseshoe-shaped bed with a keyhole entrance to limit pathways and increase planting space. The minimal pathway reduces soil compaction as less of the soil is trod upon.

The keyhole design doesn't have to be a specific shape as long as you maintain the idea that you should be able to reach each part of the garden bed from either inside the keyhole or from outside the bed.

● ● ● ● ● ●

MATERIALS

- Garden hose or flour
- Garden edging such as wood, bricks, or stone
- Cardboard
- Soil
- Mulch

MAKE IT!

1 Lay out the shape of the bed that you want to make using a garden hose or flour. If you're installing the bed over plants such as lawn or weeds, use sheet mulching to set up the foundation. Start by covering the space with cardboard and then wetting it.

2 Next add a border with bricks, wood, stones, or any other material to hold the shape. Add at least 12 inches (30 cm) of compost and garden soil. The keyhole bed can be planted right away into the compost.

3 Add a path to the center from the easiest access point. In this case, the path begins at the base of the stairs coming from the house.

4 Add 2 to 4 inches (5 to 10 cm) of mulch to the top of the bed around the plants.

Include Native Plants for Wildlife Medicine and Food

Echinacea purpurea, native to parts of Canada and the United States, feeds butterflies and birds and is used to boost the immune system.

> ◉ Include native plants in your garden.
>
> ◉◉ Design gardens with 30 to 50 percent of native plants.
>
> ◉◉ Learn the ecology and history of native plants in your area and spread the message to others.

There's a trend in home gardens these days to include native plants in home gardens. Native plants provide food, shelter, and medicine for native wildlife, and the more we select cultivated plants from our garden centers over native plants, we risk not providing the food and medicine sources that native wildlife need.

There are many benefits to growing native plants that are adapted to grow in our home garden climates. Native plants work together to support the soil and use fewer resources. They often need little supplemental watering or fertilizers, and they provide a source of seeds to populate natural spaces with native species. Native plants also encourage education and stewardship of our ecological heritage.

Now, before you head out with a trowel in hand, it's important not to remove native plants from wild spaces to plant in your garden. Removing native plants from wild spaces changes the ecology of those wild spaces, and just imagine if every gardener went out to the woods to collect all of their garden plants! It's better to look for ethical sources of native plants locally and request them to be stocked in garden centers. As more home gardeners request native plants from nurseries, they will be encouraged to increase supply, thus introducing them to many more gardeners along the way.

All of this being said, it doesn't mean that we need to exclusively plant native plants in our home gardens. There's plenty of benefit of planting edible crops, flowers, trees, and shrubs that aren't native to our area but are otherwise well suited to the climate. These plants allow us to have a wider variety of food and medicines available to us in our home gardens, plus gorgeous ornamentals, and useful plants that we can make into tools, rope, or structures such as trellises. A good start is to reserve one-third of your home garden for native plants. The percentage you choose to dedicate can be based on many factors, but don't worry about the number and just start with something. The more you learn and create a demand for them, the more we can encourage native plants as beautiful, fruitful, and historically rich garden plants.

Integrated Pest Management

It wasn't until I zoomed in on this photo of a honeysuckle bloom that I saw pest management at work in my garden.

- Accept pests as part of the ecosystem.
- Attract predators to control the pest population.
- Plant a polyculture garden to build pest resilience.

The question I probably get asked the most through my website or when I present live talks is how I deal with (*insert pest name*) in my garden. Regardless of what the pest is, I always have the same answer: find out what it's eating and what eats it. Integrated pest management in the home garden means spending a little bit of time understanding the ecology of the pest, the host plant, and the environment that's allowing that pest population to get to a stage where it's caused you some concern.

The first step is usually a reality check. Are you noticing a few pests on your plants, but they aren't causing much damage? Or are you noticing complete destruction of the host plants because the population is so out of control? Is this a pest that is known to be invasive and on a watch list in your local area? If the pest isn't on a local watch list, and it's not causing a great amount of damage, then you are probably safe to just leave it be. Congratulations, your garden is part of the ecosystem!

The only way to control pest populations in our gardens is to have a food source for predators. Predatory insects, mammals, amphibians, reptiles, and birds need those

Pests are to be expected and welcomed into our gardens. Yes, welcomed. Because without pests what would predators have to eat?

When it comes to integrated pest management just remember you don't have an aphid problem, you have a ladybug scarcity.

pests in order to survive. If we spend our time using chemical sprays to eliminate all of the pests, we're also eliminating the solution.

If your pest population is becoming out of control and causing a great deal of damage to your plants, then the next step is to ask yourself, "How do I attract the predators to my garden in order to help me control this population?" There are plenty of books and resources online where you can find specifics about each individual pest, as well as the predators that help to keep it under control. Figure those out and find the way to attract them or introduce beneficial insects, birds, and other wildlife to your garden.

You may also consider changing the host plants. If a specific plant is so overrun with pests but there really aren't any predators to control the population, then you might be better off to remove the *plant* and replace it with something that is less susceptible to attracting pests.

All in all, integrated pest management is a simple concept for a very diverse set of garden challenges. The more you introduce polyculture into your garden, the fewer problems you will have of being overrun by a particular pest. If you have multiple varieties and species of plants, there are sure to be ones that are less attractive or more resilient to the pests. Alternatively, the diverse plants could attract enough predatory insects to help you control the population.

My best advice is simply to take a walk in your garden every morning. It's a therapeutic exercise that allows you to check in on the plants on a daily basis and evaluate how they are doing. If you notice the signs of pests one day, and a few more the next, take some time to do your research and figure out what's missing in your garden.

Worm Hotel

A worm hotel can be added to bed gardens, raised beds, wicking beds, keyhole gardens, herb spirals, and even containers.

- Add a worm hotel to the garden.
- Add multiple worm hotels around the garden.
- Divide worm population to give excess to friends and neighbors.

A worm hotel is a combination of spot composting and a vermicompost bin that's buried directly into the soil. A project like this is easy enough to make multiples of for the garden, and it's ideal for small spaces. It helps to add another layer to composting around the

A worm hotel is essentially a vermicompost bin that is dug into the soil. The nutrient-rich vermicompost will be readily available to amend the soil.

garden by creating a space where worms can break down organic matter such as carbon sources in order to make nutritious worm castings to fertilize the garden.

You can make a worm hotel using any sort of tube or bucket structure as long as it's open to the soil at the bottom, and you have a lid or cap to top off the hotel. The best location for a worm hotel is an easy-to-access, high-use place in the garden where the soil could use some extra nutrients.

✿ ✿ ✿ ✿ ✿ ✿

MATERIALS
- 6- to 8-inch (15.2 to 20.3 cm) PVC tube with cap 18 inches (45.7 cm) long—for a larger garden, this project can be made with a 5-gallon bucket.
- Drill
- Carbon sources such as dried leaves, shredded paper, or straw
- Raw vegetarian kitchen or garden waste
- Finished compost
- Vermicompost worms

MAKE IT!

1 Use the drill to make holes in the bottom 6 inches (15 cm) of the PVC pipe.

2 Dig a hole in the soil that is 15 inches (38 cm) deep and insert the pipe with the drainage holes at the bottom. Backfill the soil around the pipe so that only the top 3 inches (7.5 cm) are above the ground.

3 Add two handfuls of bedding (carbon) material such as dried leaves, shredded paper, or straw to the bottom of the hotel.

4 Next add a scoop of finished compost. Water the carbon materials and the compost until it's saturated.

5 Add a handful of 150 to 200 of your favorite composting worms.

ETHICS

6 Top the worms with a 1-gallon (4.5 L) bucket of raw, vegetarian kitchen waste or non-woody garden clippings.

7 Top with a final layer of bedding material.

8 Water the inside of the worm hotel and put the cap on. Plant nutrient-hungry plants around the worm hotel.

As the worms create castings it will move out through the holes in the bottom side inches of the tube. Check to see the level of the compost and add more green waste or shredded paper as needed.

CARING FOR WORMS

Worms control their own population depending upon the space and amount of food available. If food is scarce, they will begin to eat one another and reduce their numbers. If there's plenty of food, they will continue to multiply until there are enough worms to digest the material that's been given to them. If composting worms are not happy in the soil, they will not leave the bin to forage elsewhere; they prefer to stay in the hotel. If you're feeding them well and they are happy, they may need to be divided at some point. Remove a handful of worms to populate more worm hotels or give them to a friend.

Worms are raw food, gluten-free, vegans. Feed them a variety of kitchen and garden scraps but avoid meat and dairy as well as processed and cooked foods.

TYPES OF COMPOSTING WORMS

Red Wiggler / Tiger Worm (*Eisenia fetida*)

European Nightcrawler (*Eisenia hortensis*)

African Nightcrawler (*Eudrilus eugeniae*)

Include Garden Livestock

- Introduce animals and insects into the garden to help balance the ecosystem.

- Include animals for their products and functions in the garden.

- Provide for animal and insect needs from the garden.

In our home gardens we don't often think of keeping livestock. Yet including animals and insects in our garden space is a great way to help balance the ecosystem. Birds, bees, and mammals are often kept as garden helpers and they provide a valuable service in the regenerative garden.

BEES

Many folks have turned to beekeeping not just for honey and wax production, but to help increase pollination. Bees will travel great distances and help entire neighborhoods with pollination, so if you haven't included beehives in your garden plan, hope that a neighbor does. And then thank them for all the hard work their livestock does in your garden.

BIRDS

Chickens, ducks, and quail are becoming more and more common in garden spaces. Chickens, of course, are the most common, with many urban school projects and chicken rental companies encouraging city families to include chickens in their yards. Ducks are more suited for larger properties and acreages that have ponds for them to paddle in. Quail are delightful livestock pets that produce tiny spotted eggs and have the sweetest purring and chirping sounds. Quail may not be overly practical in terms of egg production, but you will certainly get a few eggs and these birds are easy to keep in small spaces. The best thing about all of these feathered livestock is that they help to keep pest populations down and produce nutrient-rich fertilizer for your garden.

THE REGENERATIVE GARDEN

132

CATS AND DOGS

We may not consider housepets as garden livestock, but they can serve a very useful purpose in our gardens. Because cats and dogs are territorial, they will mark scent around a property with their urine, announcing to wild pests such as rodents to stay away. Cities are often overrun with rodents such as rats and mice, which can be a great challenge in the garden if their populations get out of control, especially in cities where there are fewer predators. In order to repel garden-munching wildlife such as rabbits and deer, farm supply stores will sell cougar or coyote urine that can be applied around your property. While this is effective, it's time consuming, expensive, and begs the question of how this urine is collected. A better option is to allow your house pets also to be garden pets. Allowing cats and

dogs to mark the property with urine gives you a regular natural source of predator scent to keep pests away.

PROTECTING BIRDS, LIZARDS, AND OTHER GARDEN FRIENDS FROM PETS

Cats and some dogs are natural hunters, and while that can be beneficial for scaring off pests, it can be deadly to native songbirds and lizards. Pets may also catch and eat rodents, which may be less of a concern for wildlife protection, but it could make your pet sick if they eat a poisoned or diseased rat.

Allowing pets to be part of the garden ecosystem means it's your responsibility to create a safe environment for them and the native wildlife. Keep pets safely in your garden where you can monitor them, and please spay or neuter them. You can also get effective collars [13] that cats can wear to protect birds. Bells help alert birds and rodents of a cat's presence, and colorful fabric collars have been proven effective at letting cats enjoy the outdoors without harming wild birds.

SPOT COMPOSTING

Spot composting is a very easy method of composting. Simply dig holes in various spots of the soil and add organic material into the hole. Cover it up again and allow the organic material to break down naturally over time.

WOOD SLAT BIN

Wood slat compost bins are a structured version of the compost pile. They are built directly on the ground, which allows them to bring organisms from the soil into the compost pile. Wood slats allow lots of airflow. One can easily build a wood slat bin with some basic carpentry skills, tools, and lumber. A good size for the bin is 3 feet (1 meter) wide by 3 feet (1 meter) deep by 3 feet (1 meter) tall. The slats of the back and the two sides should reach the 3-foot-tall mark. The front of the bin can be completely open or the slats can be built up halfway to allow easy filling and compost removal.

- Include compost systems that work best for your garden space.

- Include multiple compost systems.

- Produce all the compost you need for your garden from compost systems.

COMPOST PILE

A compost pile is the simplest compost system to build. Simply mound organic material in a pile and leave it. This can be as simple as piling up fall leaves and letting them compost in place. Improve this system by layering two-thirds carbon and one-third nitrogen materials in the pile and occasionally turn it to introduce air. Compost piles can be made on any kind of property but tend to look a little messy, and they need to be left in place for a long time. Covering the pile with cardboard or mulch will allow it to heat up more, but it does need to have supplemental moisture in order to stay in balance.

THE REGENERATIVE GARDEN

CRITTER-PROOF BIN

A critter-proof bin is a wood slat bin that has had critter-deterring wire mesh added. Before installing the compost bin put a thin layer of wire mesh at the bottom that comes out through the perimeter to keep rodents from burrowing in. Build all four sides all the way up and add wire mesh to the inside of the slats. Build a lid with the same structure of wooden slats and wire mesh to go on top of the bin. You can also add a rat- or mousetrap under a milk crate in the compost bin.

MULTIBIN SYSTEM

Build a multibin system by putting two or three piles or wood slat bins side-by-side. The first bin is for layering raw materials as you are out working in the garden. As that bin becomes full and the materials begin to break down, the working compost is then moved to the second bin where it is no longer added to and is allowed to finish the composting process. The finished compost can then be used in the garden, or it can be stored in a third bin.

BEAR-PROOF BIN

A bear-proof composter is made of a stone structure and a metal top to prevent bears from digging around in the compost. It can be built with bricks and mortar, stone, or poured concrete. The metal lid may have to be custom made to fit securely on the bin. Air vents that are too small for rodents and bears to access are added in order to allow airflow. To buy plans for building a bear-proof bin go to www.critterproofcomposting.com.

ROLLING COMPOST BIN

A rolling composter is a closed system that makes for neat composting in urban areas or visible garden spaces. The biggest challenge with this composting method is that the bin is not directly touching the ground; therefore, the native organisms are harder to introduce. That being said, I've had a rolling compost system in my urban garden for ten years and I never have any problems attracting organisms to break down the compost. It has two sections so one I can fill, and the other is full of working compost. I always leave a little bit of compost in the bin when I remove the finished compost in order to maintain a source of microorganisms. Note: The leach-ate that comes out of these bins is captured in a trough below, which the manufacturer often labels as "compost tea." However, it's not a compost tea. It's the runoff from the compost, which could be excess rainwater or the leftovers of that very juicy watermelon that was composted. It would be fine to use this leachate in your garden but there's no standardization for what the benefits would be because it is simply compost runoff.

Single Stem
.25 each

6
Community
Building Sharing Spaces for Everyone

Create a welcoming garden for your community members.

Build gardens in public spaces to enhance community.

Create sharing spaces that give back to the community.

The world seems to have gotten a lot smaller given our access to technology. While we are now communicating to the far reaches of the globe with ease, our view of what is directly in front of us has diminished. With so much of our attention going outward into the global world, it can be a radical act to focus our energy on the small space that is home to our garden. And yet it's increasingly important to focus our efforts at home. The soil and plants we cultivate have an impact on those that live and share the garden with us, our community.

I originally called this chapter "Wildlife" to include projects where we can ethically and responsibly host the wild creatures that call our gardens home. But that's only part of the picture. When creating a home ecosystem, our gardens become a community and a sharing space for this wildlife, yes. But the space is also shared by you, your family, pets, livestock, your neighbors, and those who pass by your home. When I thought about with whom we share garden spaces, it seemed more relevant to call this chapter "Community," which more accurately describes a garden as a living ecosystem that supports many lives.

The projects in this chapter help support this idea of community. They are inviting and attractive, so as to welcome visitors to enjoy the space. There is a bug hotel that allows insects to overwinter, a wildlife pond, pollinator plants, a butterfly rest stop, and a living hedge all designed to encourage wildlife to safely feed and nest within the garden space. There are also projects that reach out to the community such as an urban flower stand, a sharing library, a kindness victory garden, and a sensory garden for children. These projects allow us to create spaces to deepen our relationships with members of the community around us, completing the circle that supports life and growth and nourishes our families.

By creating spaces for communities, we are feeding our neighborhoods and designing places that spread joy, health, safety, and pride. Welcoming backyard wildlife helps manage pest populations, pollinate plants, and distribute seed. Welcoming neighbors helps build a culture of friendship and sharing, reduces waste, and increases connection. Once these seeds are planted, they spread and grow like wildflowers making our immediate community more joyful, lively, and healthy.

Gardening with community in mind means inviting all creatures to enjoy and share in our space.

A wildlife hedge helps to create space for biodiversity at the edge of two ecosystems.

- ● Plant a wildlife hedge at the edge of a property or where two ecosystems meet.

- ●● Mix the hedgerow plantings to stagger leafing, blooming, and fruiting times.

- Use native plants and allow the hedge to naturalize.

The ecological concept of the "edge effect" describes how there is maximum diversity where two ecosystems meet because the inhabitants of the biological communities are both present in that space. As home gardeners, particularly when a property is on the edge of a wild space or different ecosystem, creating a wildlife hedge allows space for the many creatures that share those spaces to gain protection, food, shelter, and nesting areas.

COMMUNITY

139

A wildlife hedge resembles the classic hedgerow: a living fence intended to divide property, reduce erosion, or maintain a barrier to livestock. In smaller urban and suburban home gardens, a wildlife hedge could also look like a garden border that lines fences or defines the back of a garden space. Tailoring the size of a hedge to your space will also work with the local wildlife as larger properties should have larger and more numbers of visitors, while in urban gardens the wildlife will be smaller in size and fewer in numbers.

Include a mixture of small flowering trees, shrubs, evergreens, brambles, vines, and herbaceous plants in a wildlife hedge. A diverse mix of native plants will allow for wildlife to find food and nesting sites throughout the year.

TIPS FOR PLANTING AND MAINTAINING A WILDLIFE HEDGE

- Plants that produce berries feed birds and mammals, flowers feed and host pollinators, trees and shrubs create nesting and hiding spaces, and beneficial insects have a safe playground in many different areas.
- Plan a mix of plants that will leaf, flower, and fruit at different times of the year to extend the food sources for wildlife and provide an attractive year-round display.
- Plants with thorns provide protection from larger predators (including humans) making the hedge a safe haven for wildlife.
- Evergreen plants allow a hedge to provide for wildlife throughout the seasons.
- In smaller gardens and urban areas, native plants can thrive, so it is helpful to keep on top of pruning and management of the hedge while also being careful not to remove nests or homes. In cool climates, do not prune hedges during nesting season (spring through summer).

Wildlife Pond in a Pot

A small-scale wildlife pond can be set in a large garden pot such as a wine barrel with a pond insert.

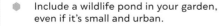 Include a wildlife pond in your garden, even if it's small and urban.

 Add native plants.

 Learn about the wildlife that visit and encourage diversity. If something becomes out of balance, look to nature for the solution.

A wildlife pond is a great opportunity for water purification and inviting biodiverse pollinators and predators to the garden. A general rule of thumb for a wildlife pond is that 30 percent of the surface area should

If the pond is going to be in a pot or raised in any way, there needs to be a ramp or climbing surface on the outside to allow wildlife to climb up and access the pond.

The plants that go in a wildlife pond can be a mix of native species and cultivated pond plants. The plantings will be both in and around the pond.

be covered with purifying plants and that the edges should be planted with reeds or tall grasses. It might also be necessary to create fencing or netting over the pond in order to prevent larger animals from accessing the pond wildlife. In some cases, ponds can be dug 6 feet (1.8 meters) deep for fish in order to allow the fish a place to escape from the prying paws of raccoons or beaks of raptors. It also provides deep water that won't freeze in winter, allowing fish to overwinter in the pond.

The wildlife that visit will vary depending on the wildlife that exists in your garden ecosystem. In an urban setting it will likely be insects and birds, possibly amphibians and lizards. Larger ponds will attract larger wildlife. The best part about creating the wildlife pond is the excitement of it becoming naturally inhabited. Wildlife are attracted to just about any water source (even puddles), so size doesn't really matter as long as you're creating a habitat that works in your garden. Wildlife ponds are most ecologically effective when they are set at the edge, the place where two or more ecosystems meet.

Floating plants act as you would expect based upon their name—they float on the surface. Their roots hang down into the water and they do not need any soil. Covering two-thirds of the surface area of the pond with floaters can help shade the pond and prevent algae from growing. Because these plants do not need to be rooted in soil, they have a propensity to multiply quickly and cover the surface of a pond. Thin them regularly so they don't cover the entire surface of a pond and smother the other inhabitants. They are nutritious food sources for some wildlife and the compost pile.

Deep-Rooted plants are also sometimes called floating plants because they have parts, like lilypads and flowers, that float on the surface. The best-known of these sun-lovers are waterlilies and lotus, which are well suited for large ponds.

Marginal plantings are grown on a step or ledge created within the pond. The plant material like leaves, stems, and flowers are above the water line and the roots are below.

Submerged plantings grow in the deeper parts of the ponds where the roots and stems are submerged, and the leaves and flowers grow up and float on the surface.

DON'T LET MOSQUITOES BUG YOU

Standing water is a perfect habitat for lots of species, including mosquitoes. If that's an insect that you don't want to invite to your habitat then you also need to encourage predators. Insects such as dragonflies, damselflies, and diving beetles love to eat mosquito larvae, as do birds, amphibians, and fish. The good news is that if you do get some mosquito larvae then you're providing a food source for some wonderful beneficial wildlife to move in. If they don't come naturally, you can always introduce a few fish to the pond (keeping in mind that these fish might also attract their own predators). A wildlife pond is meant to become an ecosystem, so as it grows in your garden you'll see different wildlife populations bloom and die back until you find the balance that is right for your garden.

● ● ● ● ● ●

MATERIALS
- Half-barrel planter
- Half-barrel pond insert or pond liner
- River stones or gravel
- Bricks, pavers, or clay pots
- Potted pond plants (or pond plants, pond soil, gravel, and pond plant baskets)

MAKE IT!

1 Select a location for the wildlife pond that is in full sun and not under trees to prevent leaves and debris from falling in. If you do choose a location under a tree, cover the pond with netting when the leaves drop or be sure to scoop out the leaves on a regular basis.

2 Level the base of the site and compact the soil.

3 Level the barrel on the base and add the insert or pond lining to make the inside watertight.

4 Place the barrel somewhere against a raised structure so that wildlife can climb up to the pond. If you are setting the pond away from structures, build a ramp to enter the pond by creating a rock wall, filled with soil and plants.

5 Build an inner area for marginal plants by stacking bricks, pavers, or clay pots turned upside down inside the wildlife pond against the side that has the wildlife ramp. Add rocks or gravel to the bottom of the pond.

6 Add two to six pond plants, which can be purchased in pots or as bare-root plants. Bare-root plants can be planted directly into the pond in which case, add a layer of pond soil mix to the planting areas before adding the rocks or gravel. Or they can be potted in mesh baskets, using pond soil mix topped with rocks or gravel. Potted plants make it easier to move when cleaning is needed, but they also require larger pots eventually (just like all container plants). Plants directly in the pond will naturalize more easily but will be harder to remove for cleaning.

7 Weight the plants with stones or gravel as needed.

8 Fill the pond with water; rainwater is best but tap water can also work. Before adding plants, let the tap water sit for 24 hours to come to air temperature and allow the chlorine to evaporate.

In areas with cold winters, the plants can be left in place over winter as long as the whole pot doesn't freeze solid. In other locales, avoid this by moving the pond to a warmer location or wrapping the outside of the pond with insulation such as burlap or a gentle heat source such as Christmas lights.

POND PLANTS LIST

As with all plants, be sure to check which are considered invasive in your area. Some plants are illegal in some areas due to the potential for them to invade waterways. Many pond plants can be beneficial to wildlife in one area yet toxic in another, so it's worth taking the time to look up each plant you want to include in your pond, even those commonly sold at garden centers. While pond plants can have a tendency to grow quickly, they are often rich in nitrogen and can safely be composted in dry land gardens.

FLOATERS

Water Lettuce	*Pistia stratiotes*
Water Hyacinth	*Eichhornia crassipes*
Duckweed	*Lemna minor*
Frogbit	*Hydrocharis morsus-ranae*
Sensitive Plant	*Neptunia oleracea*

DEEP-ROOTED PLANTS

White Waterlily	*Nymphaea odorata*
Yellow Lotus	*Nelumbo lutea*
Lotus	*Nelumbo*
Nardoo/Upright Water Clover	*Marsilea mutica*
Golden Club	*Orontium aquaticum*
Water Poppy	*Hydrocleys nymphoides*

MARGINAL OR EDGE PLANTS

Pickerelweed	*Pontederia cordata*
Chilean Rhubarb	*Gunnera tinctoria*
Various Water Iris	*Iris* spp.
Swamp Rosemallow	*Hibiscus moscheutos*
Blue Moneywort	*Lindernia grandiflora*
Yerba Mansa	*Anemopsis californica*
Swamp Verbena	*Verbena hastata*

SUBMERGENT PLANTS

Wild Celery or Eelgrass	*Vallisneria americana*
Pondweed	*Potamogeton* spp.
Water Stargrass	*Heteranthera dubia*
Mare's Tail	*Hippuris vulgaris*
Red-Stemmed Parrot Feather	*Myriophyllum brasiliensis* 'Red Stem'

EDIBLE AQUATIC PLANTS

Wild Rice	*Zizania*
Cattail	*Typha latifolia*
Water Chestnut	*Eleocharis dulcis*
Watercress	*Nasturtium officinale*
Duck Potato	*Sagittaria latifolia*

Seed-Sharing Library

A seed library is a little free library for seed sharing instead of books.

- Create a public space for seed-sharing within the community.
- Include teaching materials such as books or printed tip cards to help educate new gardeners.
- Encourage community members to collect seeds from their gardens to share.

A seed-sharing library provides community members with easily accessible seeds, seedlings, plants, cuttings, and seed-starting supplies for free. These boxes do not need to be large, but they provide great benefit. Seed libraries are an opportunity to share skills and knowledge with neighbors, bring greater

THE REGENERATIVE GARDEN

146

Most seed libraries will have a door, although in this case the seed library is only set out on days when there is good weather so the seeds are stored in the drawers. Mount the library in a shady spot so that the inside doesn't heat up too much.

food security, and provide environmental benefits. In some areas the seed library switches to a little free book library in winter that focuses on gardening and nature.

The key component to a seed library is to manage the space and engage the community members. The library structure could be as simple as a plastic container set on a post or a creative building project such as a Little Free Library. Found and recycled materials such as cabinets or public newspaper boxes can make a perfect home for a seed library. As the structure will be outdoors, it's important to consider protecting it from the elements. Adding weatherproof roofing at a slope so that water runs off will keep the inside dry. Use outdoor paint to protect the walls as well.

Include information on how to plant seeds, tend plants, and save seeds. Stock the shelves with empty seed or coin envelopes and pens. Seed envelopes can also be used to divide seed packages into smaller portions so everyone takes only what they need.

Seeds can be donated by local seed companies, with stock that they need to redistribute to the community before it expires.

This is a great opportunity for education! Community members should expect that germination rates will not be as good as they would be if the seed packages had been recently collected and stored in a cool, dry location. The seed-sharing library will not be an ideal climate for long-term storage, plus many seeds will come from seed companies that are donating old stock and community members may donate seeds past their prime. Encouraging the community to keep the seeds flowing by giving and receiving will refresh the supply enough to circulate plenty of seeds—even if germination rates are low.

There are many designs available online and in the book *Little Free Libraries and Tiny Sheds* by Phillip Schmidt and the Little Free Library. However you decide to create the box, it should be available in a public space such as a front yard or community garden and managed by a person or group.

In my neighborhood, the Zucchini Racer committee donates zucchini seeds to all the community libraries so that community members can grow their own racer for the Fall Fair.

This community flower stand was created by Holly Rodgers, an urban flower farmer who offers a small local CSA (Community Supported Agriculture) subscription. Holly sets out the flower stands in order to raise money for local charities.

- Build an urban flower stand to sell or giveaway cut flowers.

- Collect money for local charities from the sale of flowers.

- Grow enough cut flowers to have the stand open year-round.

There's nothing more heartwarming than a gardener sharing their bounty. More than once, I have driven by both urban and rural flower stands where folks can pick up a fresh bouquet from a garden overflowing with blooms. It's such a lovely sentiment.

COMMUNITY

This urban flower stand was made using a rustic wood desk built by a local woodworking shop. The desk was then customized with string to hold it open and to add signage in little wooden frames that hang from the top of the desk.

When filled in, the signs will tell passersby that fresh-cut flowers are available by donation, and to what charity the donations will be going. At the top of the desk is a permanently attached vintage wood box that is secured with a tiny lock. In the top of the box, there is a slit for donations to be deposited.

MAKE IT!

1 To make your own flower stand look for a small desk that closes with a hinged top.

2 Attach a few small wooden frames to the inside top of the desk with string or twine.

3 Print or handwrite the instructions to fit into the frames.

4 Create a donation box with a lock on it with a slit in its top for donations to be deposited. Use screws to permanently attach the donation box to the top of the desk lid.

5 Seal the wood desk with a waterproof coating to protect it from light drizzles. Keep the stand inside overnight and on rainy days. A small flower stand is portable enough that it can be brought out when there is a bounty of flowers to share and kept inside on those other days.

6 Flowers are displayed in florist buckets and Mason jars. Encourage community members to return the containers so the host can continue to supply the water-holding vases to display the flowers.

Holly also created miniature wreaths with natural vines and attached dried flowers and fresh greenery for the holidays.

FOUR SEASONS OF FLOWERS

Cut flowers can be added at any time of the year in various arrangements. Spring bulbs such as daffodils, tulips, and hyacinth will brighten up the early days of spring. Summer-flowering bulbs, annuals, and perennials make for impressive summer arrangements. In fall, there will still be plenty of flowers to cut but also include autumn-themed accompaniments such as seedheads, tiny pumpkins, and pinecones. In winter, set out festive arrangements of cut evergreens with berries along with any seasonal blooms.

KINDNESS VICTORY GARDEN

These days, we could all use a little more kindness. Surprising loved ones and neighbors with a homegrown cut flower bouquet is a small gesture that can truly brighten someone's day. When planning your regenerative garden, including a kindness victory garden—a garden specifically planted with cut flowers to give as gifts—will plant seeds of kindness and grow community joy. Here are some wonderful cutting flowers to grow in your garden that can make up many beautiful bouquets.

It's hard to know exactly how much the person who picks up the bouquet needs it, but I'm betting that it makes much more of a difference than one would expect.

It's not just the recipient of a bouquet who gets the benefit of growing kindness and love in this way. The feeling you get when you give a gift that someone truly loves, and which makes a difference in their life, is a great reward in itself. And, of course, let's not forget the true beauty and healing that comes from growing a garden, cutting and arranging the flowers, and presenting them as a gift.

ANNUALS

Snapdragon	*Antirrhinum majus*
Sunflower	*Helianthus*
Bee Balm	*Monarda*
Zinnia	*Zinnia elegans*
Cosmos	*Cosmos bipinnatus*
Cockscomb	*Celosia*
Poppy	*Papaver*
Sweet Peas	*Lathyrus odoratus*

PERENNIALS

Larkspur	*Delphinium*
Purple Coneflower	*Echinacea*
Globe Thistle	*Echinops sphaerocephalus*
Hydrangea	*Hydrangea*
Iris	*Iris*
Lavender	*Lavandula*
Peony	*Paeonia*
Rose	*Rosa*

BULBS

Ornamental Onion	*Allium*
Daffodil	*Narcissus*
Tulip	*Tulipa*
Hyacinth	*Hyacinthus*
Anemone	*Anemone coronaria*
Calla Lily	*Zantedeschia aethiopica*
Dahlia	*Dahlia pinnata*
Freesia	*Freesia*
Gladiola	*Gladiolus*
Ranunculus	*Ranunculus*
Lily	*Lilium*

City gardeners who end up with extra potatoes or salad greens are generously offering up their bounty to the community and it's just a beautiful sight to see.

THE REGENERATIVE GARDEN

- Set up a community farm stand, as permissible by local bylaws.
- Collect money and supplies for local charities.
- Maintain the farm stand throughout the year.

As gardeners, we often have a glut of edible foods such as zucchini, tomatoes, and apples ready for harvest all at once, so why not build a community farm stand to share the bounty? Community farm stands are often seen in rural communities along the roadside, but they are now popping up in urban areas as well.

Anytime you're selling food products there may be regulations in your area that should be researched. It may seem like a good idea to put out an extra dozen eggs or a basket of freshly harvested plums from your tree but there could be bylaws that restrict local food supply and distribution. That being said, most communities will accept small-scale community sharing as long as it's intended to be an unlicensed business. The neighborhood connection that results from a local farm stand is worth the efforts to change the bylaws or get an exception if it is prohibited.

Urban farm stands can take many forms depending on what folks have to share. Regardless of the offerings, they are a community-building project that provides nourishing, homegrown food to community members.

Baskets and bowls can be used to display freshly harvested goodies.

1 To build a community farm stand, you simply need a cabinet, desk, or a table and a few baskets to hold the harvest. Include a donation box if you intend to collect money or other items to help fund production or to donate to charities.

2 With the rise of freecycle and buy nothing groups, community sharing is both welcomed and deeply supported in many neighborhoods. It is a lovely project that can bring people together and encourage pride in place.

3 It also provides an educational opportunity for community members to gather and talk about growing food in the neighborhood. In the early season, there could be seed and seedling sharing. Later in the growing season, there's the opportunity for bags of harvest and microgreens and herbs. As more food flourishes, the offerings can be whatever is left after feeding the family.

Provide paper bags for folks to pack up the goodies in.

Potted plants, seedlings, seeds, and cuttings are always appreciated by farm stand customers.

Butterfly Pathway Gardens

A butterfly pathway garden doesn't need to be a specific design or include complicated plantings. Start by adding a few flowers that attract butterflies to your front garden.

- Plant butterfly hosting and food plants in your garden.

- Encourage neighbors to also plant butterfly hosting plants.

- Work with organizations that create butterfly migration pathways to get your neighborhood involved in planting paths for migrating butterflies.

Some wild creatures, such as butterflies, need to migrate in order to survive throughout the year. Unfortunately, as we've urbanized areas, we've taken away the native plants that these wonderful pollinators need in order to survive that migration. The good news is that we can create pathways along which butterflies and other pollinators make stops to help regain their energy and we can host them until they are ready to continue the journey.

Creating a station in your yard that's packed with plants that allow butterflies both to feed and reproduce is a great way to contribute to natural migration patterns. This can also be a fun community-building project to encourage neighbors to join by planting their own butterfly pathway gardens. If your community becomes passionate about it, you can reach out to other communities to join in, host workshops on how to get started, and join projects that map out migration pathways.

Many butterflies require specific plants at some point in their life cycle, so research the native butterflies that migrate in your area and learn what plants to include. For instance, monarch butterflies need milkweeds, so if monarchs are the butterflies that you're hoping to support, then your butterfly migration garden will need to include milkweed plants.

One of the key components of a planting a butterfly pathway garden is to add some signage that allows folks to know that the reason why it's there. This educational component helps others in the community to protect it and hopefully join the project themselves.

BUTTERFLY ATTRACTING PLANTS

Allium	*Allium*	Phlox	*Phlox x arendsii*
Aster	*Aster*	Poppy	*Papaver*
Bee Balm	*Monarda*	Purple Coneflower	*Echinacea*
Black-eyed Susan	*Rudbeckia hirta*	Rock Cress	*Arabis*
Blanket Flower	*Gaillardia*	Sage	*Salvia officinalis*
Butterfly Bush	*Buddleja*	Sea Holly	*Eryngium*
Catmint	*Nepeta racemosa*	Shasta Daisy	*Leucanthemum × superbum*
California Lilac	*Ceanothus*	Snapdragon	*Antirrhinum*
Cornflower	*Centaurea*	Sweet Alyssum	*Lobularia maritima*
Daylily	*Hemerocallis*	Yarrow	*Achillea millefolium*
Dill	*Anethum graveolens*	Zinnia	*Zinnia*
False Indigo	*Baptisia*		
Goldenrod	*Solidago*		
Hollyhock	*Alcea*		
Lantana	*Lantana*		
Lavender	*Lavandula*		
Lilac	*Syringa vulgaris*		
Lupine	*Lupinus x hybrida*		
Milkweed	*Asclepias*		
Nasturtium	*Tropaeolum*		

Hopscotch steppingstones encourage little feet to walk through the garden.

- Plant a garden in your front yard or a public space for children to enjoy.

- Include plants that help stimulate the five senses.

- Invite children to interact with the garden, answer their questions, and allow them to help with garden chores like planting or harvesting.

In my first home, I planted plenty of plants in my front garden in order to attract the urban wildlife: neighborhood kids! I was lucky to live on a street with plenty of children, so I added plants specifically for the kiddos who passed by. Years later when the children were older, they shared with me how meaningful my garden had been to them. I truly believe that teaching children to love the earth encourages them to protect it.

COMMUNITY

159

Unlike adults, who are so familiar with everything around us that we no longer recognize the small or common things, children's eyes, ears, tongues, noses, and fingers are wide open to all the brand-new sensations of this world.

Sight: Children can see the overall beauty that grows in the garden, but they more often hone-in on smaller details and really inspect them. Plant flowers and grasses down on the ground along with tall lilies and alliums that tower above them. Tiny decorative blooms, bright-colored berries, and funny shapes stimulate delight and wonder. To engage children more, allow them pick flowers and pull them apart to get a closer look with a magnifying glass. Collect insects in a jar and discuss what they might like to eat or what habitat they need to survive. Play "I Spy" and see how many new things they can observe in the outdoor space.

A pot for digging.

Through developing gardens that allow children to see, touch, taste, smell, and hear the plants and wildlife in a garden, they are meeting plants in a way that will create a long-lasting relationship.

'Pink Lemonade' blueberries are a fun children garden berry to plant.

Smell: Try having children smell different flowers to see how they differ from one another, or smell different flower colors of the same species, and write down the results. My child is convinced that all tulips smell like a fruit that is the same color, although I have yet to smell the aroma of banana from yellow tulips.

Taste: Eating berries fresh from the garden stimulates joy and connection with the sweetness of plants. Plant blueberries, grapes, strawberries, raspberries, and ground cherries all for the simple pleasure of snacking out in the garden. Don't forget the vegetables and flowers. Peas, beans, violas, and especially kale flowers are favorites. Herb leaves offer a fun tasting activity; pick a few different kinds of herbs and let little ones try to identify them just from their flavor. Get ready for some funny faces when chives or rosemary get chomped. Their palates will grow the more they snack.

Touch: Feel the soft leaves of lamb's ear, the spiky casing of a chestnut, the warmth of river stones set in the sun, or the squishy guts of a freshly harvested tomato. There is no end to what little hands can get into in the garden so plan the garden to be hardy enough to be picked, trampled, and touched. Then discuss how each flower that's picked removes pollinator food and won't become fruit. Children will learn the life cycles and protect those delicate blooms next time.

Sound: Ask children to identify the sounds they hear in the garden. Is there running water, buzzing bees, or chirping birds? How many different things make up the musical score that plays in the background? Challenge children to imitate the sounds and find the source.

FUN VEGETABLES FOR A CHILDREN'S GARDEN

'Aunt Molly's' Ground Cherries	*Physalis pruinosa*	Tasty pineapple-citrus-butterscotch-flavored fruit that grow close to the ground in husks like tomatillos. Fun to pick and tasty to eat.
'Baby Bottle' Gourd	*Lagenaria siceraria*	A miniature version of the birdhouse gourd. Average 2 inches (5 cm) in diameter and 4 inches (10 cm) long. Their small size makes them ideal for fall table arrangements and for crafting.
'Candy Cane' Pepper	*Capsicum annuum*	Eye-catching variegated foliage and fruit that ripen from green striped to solid red make this sweet pepper very unique and will make kids excited to eat their vegetables.
'Candyland Red' Tomato	*Lycopersicon esculentum*	The branched trusses of the compact currant tomato plant are easily accessible for easy harvest. Tiny, very sweet fruit.
'Chioggia' Beets	*Beta vulgaris*	This candy-striped heirloom has pink-and-white rings in the center that make it as fun to eat as it is tasty.
'Cosmic Purple' Carrots	*Daucus carota*	Not all carrots are orange! Purple carrots have a secret though—they are orange in the center.
Cucamelon	*Melothria scabra*	Oblong green-striped 1-inch-long (2.5-cm-long) fruit that look like baby watermelons but taste like a lemony cucumber.
'Dragon's Tongue' Bean	*Phaseolus vulgaris*	These stringless, crisp, and juicy snap beans turn from lime green to a buff yellow with bright purple stripes.

Cucamelons and 'Candyland Red' currant tomatoes.

'Hot Lips' Sage	*Salvia microphylla*	The little flowers on this sage look like lips with red lipstick on.
'Easter Egg' Radish	*Raphanus sativus*	Radish bunches come up in a variety of pastel colors such as white, rose pink, bubble-gum pink, amethyst, mauve, scarlet and purple. Ready to pick in just 28 days.
'Lil Pump-Ke-Mon' Pumpkin	*Cucurbita* spp.	Features orange-and-green stripes on a white background for a unique look that is great for fall decorations. The compact, space-saving vine yields plenty of fruit.
'Mammoth' Sunflower	*Helianthus giganteus*	Grow giant sunflowers in a children's garden for the sheer delight on their faces, plus the seeds are easier to harvest.
Parisian Carrot	*Daucus carota* subsp. *sativus*	Short, round, and very popular with children, this variety of ball carrot evokes pure joy upon pulling it out of the ground. It will also grow in places where other carrots won't, such as clay or rocky soil.
Pineapple Sage	*Salvia elegans*	Bright red, fragrant flowers. The crushed leaves have the aroma and taste of pineapple for a lovely iced tea.
'Sungold' Tomato	*Lycopersicon esculentum*	This delicious super-sweet tomato has great yields and is always a hit with kids.
Stevia	*Stevia rebaudiana*	A naturally sweet leaf that can be chewed in the garden or added to tea and desserts.
Violas	*Viola cornuta*	Beautiful edible flowers can be used to decorate cupcakes and pizza.
Yard-Long Beans	*Vigna unguiculata* ssp. *sesquipedalis*	These impressively long beans are fun to play with and tasty to eat.

- Create spaces for insects to overwinter.

- Research the insects that overwinter in your garden and find more ways to give them a safe space.

- Place bug hotels in neighborhood public spaces to encourage overwintering and community education.

A bug hotel is an overwintering space for a variety of native insects that thrive in your garden. Crafting a structure of natural materials that provide crevices for insects to hibernate and nest is not necessary in wild spaces where the habitat is left undisturbed. In our home gardens, we are constantly working on the garden and changing these spaces. Adding a bug hotel on a fence gives your garden insects a safe space to rest protected from garden work and visitors.

In addition, a bug hotel looks beautiful and provides a teaching space that allows children and neighbors to learn about the importance of maintaining space for insects in the garden. Every time you pass by, you will be reminded to welcome insects and other wildlife to your garden and thank them for all the hard work they do for you and your plants.

In this project, you will build a frame and adhere plant material to it to create shallow areas for insects to burrow. It will provide the space for some insects, but others will need more depth to remain protected from predators. Through leaving plant material in the garden in the fall and winter months, allowing leaves to stay where they fall or using leaves as mulch, and densely planting your plants, there will be plenty more areas for insects and a thriving ecosystem.

········

MATERIALS AND TOOLS

Makes a 2- x 2-foot (0.6- x 0.6-meter) square, 1½-inch-deep (3.5-cm-deep) bug hotel from reclaimed wood.

- Reclaimed planks such as barn board
- 2-foot x 2-foot x ¾-inch (0.6-meter x 0.6-meter x 2-cm) square plywood
- Waterproof wood glue
- 1-inch (2.5-cm) finishing nails
- Table saw or handsaw
- Miter saw or handsaw and miter box
- Tape measure
- Square
- Sticks, branches, seedheads, bamboo, moss, and more
- Powerful hand pruners

MAKE IT!

1 Saw four pieces of reclaimed wood to 25 inches (0.6 meter), 1 inch (2.5 cm) greater than the outside length of the frame height and width.

2 Using a miter saw set at a 45-degree angle, cut a miter joint on each side of the wood pieces.

3 Assemble the four pieces into a square frame on the plywood board, overhanging by ½ inch (1 cm) on all sides.

4 Apply waterproof wood glue onto each miter joint and then fasten the pieces together with a nail close to each outside corner. Allow to dry according to timing indicated on the glue.

5 Once dry, apply waterproof glue along the back side of the frame no more than ½ inch (1 cm) from the inside edge of the frame.

5

6 Nail the plywood board to the frame from the back side. Allow the glue to dry completely.

While the frame is drying, head out and collect materials.

7 Use a powerful pair of pruners that can handle the job of cutting through thick branches and begin cutting the materials to lay out a design.

8 Start with a wave of cut wood slices diagonally across the frame. Then fill in more space on the edges with cut branches. The smaller pieces such as seedheads can be added last to add more detail and decoration.

9 Glue all of the elements down by lifting them up one at a time, adding wood glue to the back, and then setting them into place. Allow the glue to dry completely. Secure the larger pieces using 1-inch (2.5-cm) finishing nails hammered in from the back.

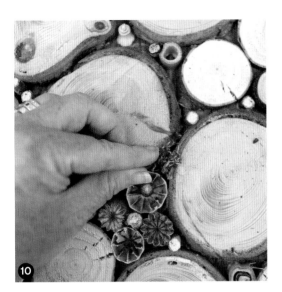

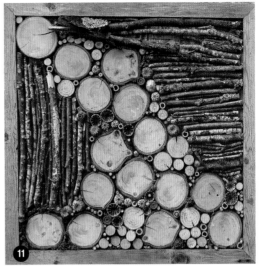

10. Complete the artwork by adding moss into any open spaces or corners that look like they could benefit from a pop of green.

11. Set the bug hotel on a fence out of direct sunlight. If it is protected from the elements, it will last longer; however, it's expected that the wood and materials will weather with time and use from nesting inhabitants. Sit back and enjoy how this art changes over time!

MATERIAL SOURCING

Look for materials from your garden or surrounding neighborhood and you will have an original piece that is truly art imitating life. Choose straight branches with lichen and mosses on them. Find some large branches that can be cut into slices. Look for wood that has interesting marks and insect holes to add a little bit of character to the art piece.

Resources

Robin Wall Kimmerer
Braiding Sweetgrass: Indigenous Wisdom,
*   Scientific Knowledge and the Teachings of Plants*
(Milkweed Editions, 2015)

Rosemary Morrow
Earth User's Guide to Permaculture
(Permanent Publications, 2010)

Michael J. Caduto, Joseph Bruchac
Native American Gardening: Stories, Projects,
*   and Recipes for Families*
(Fulcrum Publishing, 1996)

Toby Hemenway
Gaia's Garden: A Guide to Home-Scale
*   Permaculture Second Edition*
(Chelsea Green Publishing, 2009)

Oregon State University College of
    Agricultural Sciences
Department of Horticulture / Permaculture,
https://horticulture.oregonstate.edu/permaculture

Charles Dowding
Organic Gardening: The Natural No-Dig Way
(Green Books, 2013)

Delvin Solkinson with Kym Chi (editors)
Permaculture Design Notes
(Permaculture Design, 2017)
www.permaculturedesign.ca

Bill Mollison, Reny Mia Slayug
Permaculture: A Designers Manual
(Tagari, 1997)

Jessica Walliser
Plant Partners: Science-Based Companion
*   Planting Strategies for the Vegetable Garden*
(Storey Publishing, 2020)

Peter Wohlleben, Tim Flannery, et al.,
The Hidden Life of Trees: What They Feel,
*   How They Communicate—Discoveries from*
*   A Secret World*
(Greystone Books, 2016)

12,000 Rain Gardens in Puget Sound
https://www.12000raingardens.org

National Gardening Association
https://garden.org/

Jessi Bloom, Dave Boehnlein,
    Paul Kearsley (Illustrator)
Practical Permaculture: for Home Landscapes,
*   Your Community, and the Whole Earth*
(Timber Press, 2015)

Seattle RainWise
https://700milliongallons.org/tools/

Siskiyou Permaculture
https://sites.google.com/site/siskiyouper/about-us/
hazel-ward-biography

Charlie Nardozzi
The Complete Guide to No-Dig Gardening:
*   Grow beautiful vegetables, herbs, and*
*   flowers—the easy way!*
(Cool Springs Press, 2020)

Darrell Frey, Michelle Czolba
The Food Forest Handbook: Design and Manage
*   a Home-Scale Perennial Polyculture Garden*
(New Society Publishers, 2017)

Toby Hemenway
The Permaculture City: Regenerative Design
*   for Urban, Suburban, and Town Resilience*
(Chelsea Green Publishing, 2015)

The National Wildlife Federation
www.nwf.org

About the Author

AWARD-WINNING AUTHOR AND CREATOR OF GARDEN THERAPY®
GardenTherapy.ca

After becoming severely and suddenly disabled, Stephanie Rose learned to garden to heal her body and found a lifelong love for plants. More than fifteen years later, Stephanie aims to encourage joy and healing through gardening through her work as an author, permaculture designer, and herbalist in Vancouver, Canada. Stephanie loves working with the youngest gardeners and volunteers to develop children's gardens as a Vancouver Master Gardener.

Stephanie has written eleven books including *Garden Alchemy: 80 Recipes and Concoctions for Organic Fertilizers, Plant Elixirs, Potting Mixes, Pest Deterrents, and More* (Cool Springs Press, 2020); *Home Apothecary* (Leisure Arts, 2018); *The Natural Beauty Recipe Book* (Rose Garden Press, 2016); and *Garden Made: A Year of Seasonal Projects to Beautify Your Garden and Your Life* (Roost Books, 2015), which was a Gold Medal Winner from the 2016 Independent Publishers Book Awards (the "IPPYs"). She is an active member of Garden Communicators International (GardenComm), the Permaculture Institute of North America, and the International Herb Association.

Stephanie is passionate about organic gardening, natural healing, and art as part of life. Stephanie lives with her kiddo, cat, dog, and 500 worms. She shares stories, recipes, and projects on her inspiring website GardenTherapy.ca.

Acknowledgments

A book of this scale is never authored by just one person. I may be the vehicle that made the words and pictures appear on the page, but I couldn't have done any of it without the guidance, teaching, support, and love of the people who have influenced or participated in this amazing journey.

I'm grateful to have a group of passionate gardeners in my life who are building regenerative gardens. Laura Crema; James, Kristen, and Marissa from Harris Seeds; Holly Rodgers; Steven, Emma, and Quinn Biggs; Anne Marie, Trevor, Dax and Koen; Tessa, Simon, Eugenio, and Allison; Michael; Melody Kurt; and Anneli Piigert—thank you for building some of the gardens and projects that star in this book. Your passion is inspiring.

The incredibly talented Eduardo Cristo took time away from his music studio to pick up a camera and capture me in my natural element (with dirty fingernails and a smile, of course!). So much artistic love goes to Tanya Anderson, Susan Gobel, Lori Weidenhammer, Catherine Clark, and Ada Keesler for their creative contributions.

To teachers and friends Rosemary Gladstar, Charlie Nardozzi, Deborah Jones, Lori Snyder, Kym Chi, Peggy Anne Montgomery, Stacy Friedman, Jo Sullivan and the Seattle RainWise Crew, Dave Whitinger and the National Gardening Association, the Vancouver Master Gardeners, Village Vancouver Permaculture, City Farmer, the Yarrow Ecovillage, and the Cougar Creek Streamkeepers—I'm grateful for your selfless sharing of wisdom and space.

To Jessica, Steve, Anne, John, and the whole team at Cool Springs Press / Quarto— I'm a better writer for making books with you. That you all believe in the work we are creating with passion and respect makes the final copy so much richer. I know that everyone who reads it feels the authenticity that comes from working with folks who truly care about what we do.

As I sit here now at my computer, sipping a celebratory glass of wine, I would like to raise my glass to all those folks who contributed to this book. Not just those whom I have mentioned, but the people who take time out of their day to send me an email or message me with your story, experiences, and thoughts. Thank you to those who take time to share the unique perspectives that help us all grow. Keep telling your story so that it may reach those who need it.

Citations

1 https://sites.google.com/site/siskiyouper/tom-s-blog-articles

2 https://www.ag.ndsu.edu/publications/livestock/composting-animal-manures-a-guide-to-the-process-and-management-of-animal-manure-compost

3 https://extension.oregonstate.edu/news/turn-manure-compost-your-garden

4 https://www.extension.umd.edu/sites/extension.umd.edu/files/_images/programs/hgic/Publications/HG42_Soil_Amendments_and_Fertilizers.pdf

5 https://extension.umd.edu/learn/gardener-alert-beware-herbicide-contaminated-compost-and-manure

6 *The Complete Guide to No-Dig Gardening: Grow beautiful vegetables, herbs, and flowers—the easy way!* By Charlie Nardozzi, December 15, 2020

7 https://covercrops.ca/lupin/

8 https://www.noble.org/videos/cover-crop-series-mung-bean/

9 Sources: Permaculture Design Notes, Delvin Solkinson and Kym Chi; BC Farms and Food: https://bcfarmsandfood.com/what-weeds-can-tell-you-about-your-garden/

10 https://www.permaculturenews.org/2010/09/16/ollas-unglazed-clay-pots-for-garden-irrigation/

11 https://www.livingwatersmart.com.au/gardenwateringneeds

12 Rain garden measurements provided by Seattle Public Utilities and King County Wastewater Treatment Division's RainWise Program, www.700milliongallons.org

13 www.birdsbesafe.com/

Index

horse manure, 17
horsetail (*Equisetum arvense*), 34
'Hot Lips' sage (*Salvia microphylla*), 163
hugelkultur, 28–31, 83
hyacinth (*Hyacinthus*), 144, 152, 153
hydrangea (*Hydrangea*), 153

I

integrated pest management, 125–126
intensive planting, 69
iris (*Iris*), 153

J

Japanese anemone (*Anemone hupehensis*), 95
Johnny jump-up (*Viola tricolor*), 71

K

keyhole beds, 121–123
kidney weed (*Dichondra micrantha*), 71

L

lantana (*Lantana*), 158
larkspur (*Delphinium*), 153
lattice-shaped espalier trees, 87
lavender (*Lavandula*)
    bees and, 95
    butterflies an, 158
    herb spirals and, 106
    victory garden and, 153
lawn alternatives, 70–71
leaf mold, mulching with, 23, 25, 26
lemon balm (*Melissa officinalis*), 75, 95
lettuce
    herb spirals and, 106
    interplanting herbs with, 64
    water lettuce, 145
lilac (*Syringa vulgaris*), 158
'Lil Pump-Ke-Mon' pumpkin (*Cucurbita* spp.), 163
lily (*Lilium*), 153
linden (*Tilia* spp.), 74
Little Free Libraries and Tiny Sheds (Phillip Schmidt), 148
livestock, 132–133, 140
living mulch, 24
llama manure, 17

lotus (*Nelumbo*), 145
lovage (*Levisticum officinale*), 74
lupin (*Lupinus* spp.)
    butterflies and, 158
    as green manure, 19, 21
    guilds and, 75

M

'Mammoth' sunflower (*Helianthus giganteus*), 163
manures. *See also* composting; green manures; mulching; soil.
    nutrients, 16
    soil amendments with, 16–17
    types, 17
mapping, 107–108
mare's tail (*Hippuris vulgaris*), 145
marigold (*Tagetes* spp.), 64, 74, 95
medicines
    guild, 75
    native plants as, 32, 124
micro clover (*Trifolium repens* var. *pipolina*), 71
milkweed (*Asclepias*), 158
millet (*Pennisetum glaucum*), 95
mimosa / silk tree (*Albizia julibrissin*), 75
miner plants, 73
mint (*Mentha* spp.), 64, 74, 95
monoculture cropping, 65
montbretia / falling stars / copper tips (*Crocosmia*), 95
morning glory (*Convolvulus arvensis*), 34
mosquitoes, 42, 52, 55, 143
mulching. *See also* composting; green manures; manures; soil.
    aquatic plants, 25
    cardboard, 24
    chop-and-drop, 25, 73
    conifer needles, 23–24
    grass clippings, 24
    hugelkultur, 28–31
    leaf mold, 23
    living mulch, 24
    pine straw, 23–24
    rocks as, 23
    seaweed, 25
    sheet mulching, 26–27
    snow as, 25
    straw, 24
    wood chips, 23
mullein (*Verbascum thapsus*), 29, 34
multibin composting systems, 135
mung bean (*Vigna radiata*), 21
mushrooms
    compost, 17

food forests, 81
mustard (*Brassica* species), 22

N

nanking cherry (*Prunus tomentosa*), 74
nardoo / upright water clover (*Marsilea mutica*), 145
Nardozzi, Charlie, 20
nasturtium (*Tropaeolum* spp.)
    bees and, 95
    butterflies and, 158
    guilds and, 74
    interplanting, 64
native plants, 124
nettles (*Urtica dioica*), 75
Noble, Kristen, 103

O

oats (*Avena sativa*), 22
oilseed radish (*Raphanus sativus*), 22
olla catchment systems, 42–45
onion (*Allium cepa*), 74, 95, 153
oregano (*Origanum vulgare*), 74, 95, 106
ornamental onion (*Allium*), 95, 153

P

pansies, 106
Parisian carrot (*Daucus carota* subsp. *sativus*), 163
parsley (*Petroselinum crispum*), 74, 106
pears (*Pyrus* spp.), 74
peas
    as green manure, 21
    seed collection, 77
    victory garden, 153
peony (*Paeonia*), 153
peppermint (*Mentha × piperita*), 75
perennial salvia (*Salvia*), 95
periwinkle (*Vinca minor*), 71
pests
    critter-proof composting bins, 135
    dogs and, 133
    interplanting and, 63, 65
    management, 73, 125–126
    mosquitoes, 42, 52, 55, 143
pets, 133
phacelia (*P. tanacetifolia*), 20, 22
phlox (*Phlox* x *arendsii*), 158